Artillery of the World

CHRISTOPHER F. FOSS

ARTILLERY OF THE WORLD

LONDON

IAN ALLAN LTD

First published 1974

ISBN 0 7110 0505 2

To my wife Elaine

Published by Ian Allan Ltd, Shepperton, Surrey, and printed in the United Kingdom by Morrison & Gibb Ltd, London and Edinburgh

Contents

Abbreviations

AA Anti-Aircraft
AE Atomic Explosive
AP Armour Piercing
APDS Armour Piercing Discarding Sabot
APDS-T Armour Piercing Discarding Sabot-Tracer
APHE Armour Piercing High Explosive
API Armour Piercing Incendiary
API-T Armour Piercing Incendiary-Tracer
ATG Anti-Tank Gun
Aux Prop Auxiliary Propelled
ECM Electronic Counter Measures
ECCM Electronic Counter Counter Measures
FCS Fire Control System
FDC Fire Direction Centre
FG Field Gun
FH Field Howitzer
GH Gun/Howitzer
gr grammes
HE High Explosive
HEAT High Explosive Anti-Tank
HEI High Explosive Incendiary
HEP-T High Explosive Plastic-Tracer
HE-S High Explosive-Spotting
HESH High Explosive Squash Head
HOW Howitzer
hr hour
HVAP High Velocity Armour Piercing
IFF Interrogation (Identification) Friend or Foe
kg kilogramme
kg/cm^2 kilogrammes per square centimetre
km kilometre
LAAG Light Anti-Aircraft Gun
LFH Light Field Howitzer
m metre/metres
mm millimetres
m/s metres/second
m/v muzzle velocity
MG Machine Gun
mph miles per hour
MRS Multiple Rocket System
MTBF Mean Time Between Failures
PH Pack Howitzer
PPI Plan Position Indicator
RAP Rocket Assisted Projectile
rd round (rounds)
R/F Range Finder
RL Rocket Launcher
ROF Rate of Fire
rpm rounds per minute
RR Recoilless Rifle
SP Self-Propelled
t tonne
TP Target Practice
w/o without
WP White Phosphorous

Introduction

This book is the first known attempt to describe and illustrate all of the current 'Artillery of the World'. Included are anti-tank guns, field guns and howitzers, anti-aircraft guns, mortars and recoilless weapons. A few weapons that are not yet in service are included, for example the British 105mm Light Gun and the American XM-204 105mm Light Howitzer.

The word 'artillery' has been taken to include recoilless rifles and mortars of 81mm and over. These, in most armies, are the domain of the infantry rather than the artillery. I have taken the line that if the weapon can be carried complete by one man then it is infantry, if not it is artillery.

Towards the end of the book are tables and photographs of tracked prime movers, artillery rockets, comparative data of self-propelled guns. The Soviets have used artillery rockets for many years, mainly with HE warheads, and Allied forces used them during World War II. Recently the West has taken an interest in Multiple Rocket Systems, primarily in the anti-tank role. The German LARS is an example.

The last part of the book contains a very small section of the ancillary equipment that has been developed over the last few years. These include fire control systems for anti-aircraft guns, computers, laser rangefinders and muzzle velocity measuring devices.

Should the reader find any errors or omissions the author would like to know. Also, should any individual, governments or companies have any additional, or fresh material that has been cleared for publication for future editions of *Artillery of the World*, it should be forwarded to '*Artillery of the World*', Ian Allan Limited, Terminal House, Shepperton, TW17 8AS, ENGLAND.

The author would like to thank the many governments, companies and individuals who have assisted in the preparation of this book. Special thanks are due to Richard M. Bennett, G. Tillotson and C. W. Moggridge for their most valuable assistance.

March 1973 **Christopher F. Foss**

Acknowledgements

The author would like to thank the following for their most valuable assistance in the preparation of this book:

British Manufacture & Research Company Limited
Contraves AG, Zurich, Switzerland
EMI Limited
Ferranti Limited (Electronic and Display Equipment Division)
Finnish Ministry of Defence
General Electric Company (Armament Department), Vermont, USA
German Army (Public Relations)
Gesellschaft Für Ungelenkte Flugkörpersysteme MBH
N.V. Hollandse Signaalapparaten
Indian Government
Israel Defence Forces (General Staff)
Litton Systems Incorporated (Data Systems Division)
Lockheed Aircraft Service Company
Marconi Space and Defence Systems Limited (Military Division)
Nissan Motor Company Limited
Norwegian Army
Oerlikon-Bührle Limited (Switzerland)
Radartronic A/S
Republic of Vietnam
Rheinmetall GmbH
Saab-Scania (Aerospace Division)
Simonsen Radio, Norway
Sociéte Européenne de Propulsion, France
Swedish Army (Army Material Department)
Thomson-Brandt (Armament Department)
Thomson-CSF, Division Radars de Surface
United States Army including:
  Headquarters, United States Army Weapons Command, Rock Island.
  Office of the Chief of Information, Department of The Army.
  United States Army Europe.
J. I. Taibo, Spain
Susuma Yamada, Japan

Communist China

37mm Type 55 Light Anti-Aircraft Gun
This is the Soviet M-1939 LAAG built in China. Some have also been supplied to Cambodia, Laos, North Vietnam, Pakistan and Tanzania.

57mm Anti-Tank Gun Type 55
This is the Soviet 57mm Anti-Tank Gun M-1943 built in China.

57mm Recoilless Rifle Type 36
This is the American 57mm M-18 Recoilless Rifle modified by the Chinese and built in China. Has also been exported to a number of countries including North Vietnam.

75mm Type 94 Mountain Pack Howitzer
This is the pre-war Japanese gun built in China, also used by North Vietnam. For full details refer to the Japanese section.

76mm Type 54 Light Field Gun
This is the Soviet M-1942 (ZIS-3) 76mm gun built in China. It has also been exported to a number of countries including Cambodia, Tanzania and North Vietnam. Full details are given in the Soviet section.

75mm Recoilless Rifle Type 52
This is a modified United States M-20 Recoilless Rifle built in China. Some have been supplied to North Vietnam.

75mm Recoilless Rifle Type 56
This is a modified type 52 weapon.

82mm Mortar Type 53
This is a modified M-1937 Soviet Mortar. Has also been exported to Cambodia, North Vietnam, Pakistan, Tanzania and Uganda. For full details refer to the Soviet section.

90mm Type 97 Mortar
This is basically a Japanese 90mm Mortar Model 97 (1937). Total weight is 105kg and range 3700m. Very similar in appearance to the American 81mm Mortar M-1. Has also been exported to Algeria.

120mm Type 53 Mortar
This is the Soviet 120mm M-1943 Mortar built in China. Has also been supplied to North Vietnam, Pakistan and Tanzania. Has also been called Type 55.

160mm Type 60 Mortar
This is reported to be the Soviet M-1953 160mm Mortar built in China.

M-1946 130mm Field Gun (Soviet Designation)
It is reported that the above is being built in China and that some have been supplied to Pakistan.

NOTE: *It is also probable that China is manufacturing other types of mortars and artillery.*

Chinese Type-55 Light Anti-Aircraft Gun used by the Tanzanian Army

Czech 152mm Howitzer M18/46 in firing position, note the shield and the wheels

152mm Howitzer M18/46 Czechoslovakia

Calibre: 152·4mm
Weight Firing: 5512kg
Length Travelling: 8284mm
Width Travelling: 2255mm
Height Travelling: 1707mm
Barrel Length: 4875mm (including muzzle brake)
Elevation: 0° +45°
Traverse: 60° (total)
Maximum Range: 12,400m (HE round)
Rate of Fire: 4rpm
Ammunition: HE wt 39·9kg, m/v 508 m/s
Semi-AP wt 51·1kg, m/v 432 m/s
Armour Penetration: 82mm at 1000m with Semi-AP round
Crew: 7
Towing Vehicle: KrAZ-214, 6×6, Truck

At the end of World War II the Czechoslovakians had a large number of the German 15cm s.F.H.18 Medium Howitzers left behind by the Germans. The Czechs rebored the weapons to fire the Soviet 152mm Howitzer round and also added a straight topped shield and a large double baffle muzzle brake.

It has a hand-operated horizontal sliding wedge type breechblock, hydraulic recoil buffer and a hydropneumatic recuperator. The carriage has split trails and solid tyred wheels. The ammunition is of the case-type, variable charge, separate-loading type.

Employment
Czechoslovakia (reserve).

105mm Howitzer M18/49 Czechoslovakia

Calibre: 105mm
Weight Firing: 1750kg
Length Travelling: 5980mm
Width Travelling: 2100mm
Height Travelling: 1800mm
Track: 1560mm
Ground Clearance: 400mm
Barrel Length: 3308mm (including muzzle brake)
Elevation: −5° +42°
Traverse: 60° (total)
Maximum Range: 12,320m (HE round)
Rate of Fire: 6–8rpm
Ammunition: HE wt 14·8kg, m/v 540 m/s
Crew: 8
Towing Vehicle: PRAGA V3S, 6×6, Truck
TATRA-111R, 6×6, Truck

Czech 105mm Howitzer M18/49 in firing position

This weapon is the German World War II Light Field Howitzer M18 mounted on the modified carriage of the German PAK.40, and known in the German Army as the leFH 18/40.

The Czechs fitted the weapon with single rubber tyres in place of the heavy spoked wheels that had solid tyres. The weapon has a double baffle muzzle brake, a hydraulic recoil buffer and a hydro-pneumatic recuperator, the breechblock is of the hand-operated horizontal sliding wedge type. The ammunition is of the case-type, · variable-charge, separate-loading type.

Employment

Czechoslovakia (reserve).

100mm M-1955 Field Anti-Tank Gun

Czechoslovakia

Calibre: 100mm
Weight Firing: 3400kg
Length Travelling: 8534mm
Width Travelling: 2179mm
Height Travelling: 2606mm
Track: 1969mm
Barrel Length: 6431mm (including muzzle brake)
Elevation: $-5° +40°$
Traverse: 60° (total)
Maximum Range: 21,000m (HE round)
Rate of Fire: 8rpm
Ammunition: APHE wt 15·9kg, m/v 1000 m/s
HE wt 15·7kg, m/v 900 m/s
HEAT round m/v 800 m/s
Armour Penetration: 185mm at 1000m (APHE round)
380m at any range (HEAT round)
Crew: 6
Towing Vehicle: TATRA-138, 6×6, Truck

This is also called the 100mm Field Gun M53 and is built in Czechoslovakia. It is the Czechoslovakian equivalent of the Soviet 100mm field gun, and fires the same ammunition as the latter and has similar performance.

It has a semi-automatic horizontal sliding wedge type breechblock and a straight top to its shield, the sides of the shield slope to the rear, long barrel with a double baffle muzzle brake, split trails and single tyres. The gun can be fitted with the APN-3-5 infra-red sight for night firing.

Employment

Czechoslovakia.

85mm M52 Field Gun.

85mm M52 Field Gun — Czechoslovakia

Calibre: 85mm
Weight Firing: 2095kg
Length Travelling: 7520mm
Width Travelling: 1980mm
Height Travelling: 1515mm
Track: 1630mm
Ground Clearance: 350mm
Elevation: $-6°$ $+38°$
Traverse: 60° (total)
Barrel Length: 5070mm (including muzzle brake)
Maximum Range: 16,160m (HE round)
Rate of Fire: 20rpm
Ammunition: HE wt 9·5kg, m/v 805 m/s
APHE wt 9·3kg, m/v 820 m/s
HVAP wt 5·0kg, m/v 1070 m/s
Armour Penetration: 123mm at 1000m APHE round
107mm at 1000m HVAP round
Crew: 7
Towing Vehicle: PRAGA V3S, 6×6, Truck
TATRA-111R, 6×6, Truck

This weapon is the Czech equivalent of the Soviet D-44 gun, it was designed and manufactured in Czechoslovakia. It fires the same type of fixed ammunition as the various Russian 85mm guns and has a similar performance. It can be mistaken for the Soviet or Czech 100mm guns. The M52 has stowage boxes on the trails.

	Soviet 100mm M-1944 ATG	**Czech 100mm FG.M-53**	**Czech 85mm M52 FG**
Breechblock	vertical sliding wedge	horizontal sliding wedge	vertical sliding wedge
Shield Top	straight	straight	wavy
Wheels	dual	single	single
Trails	box section	box section	box section
Muzzle Brake	double-baffle	double-baffle	double-baffle

Employment
East Germany, Czechoslovakia.

M59A being uncoupled from its UAZ-69 towing vehicle

82mm Recoilless Gun M59A Czechoslovakia

Calibre: 82mm
Weight (Firing and Travelling): 386kg
Length Travelling: 4280mm
Width Travelling: 1675mm
Height Travelling: 1000mm
Track: 1520mm
Ground Clearance: 325mm
Height of Axis of Bore at 0° Elevation: 805mm
Elevation: −13° +25°
Traverse at 0° Elevation: 360°
Traverse at 25° Elevation: 60°
Rate of Fire: 6rpm
Ammunition: HE wt 6kg, m/v 565 m/s
HEAT wt 6kg, m/v 745 m/s
(above weights are projectile weights)
Maximum Range: 7560m (HE round)
Point Blank Range 2m High Target: 755m (HEAT round)
Armour Penetration: 250mm (HEAT round)
Crew: 5
Towing Vehicle: TATRA-805, 4×4, Truck
UAZ-69, 4×4, Truck
OT-810, Half-Track

This weapon has the same calibre as the lighter T-21 but a much greater range. It is built by the Škoda works in Pilsen. The weapon is employed in the Czechoslovakian Army within the motorised rifle regiment as an anti-tank gun for infantry supporting fire using both direct and indirect laying.

It is normally towed by its breech end behind a truck or APC, it is also often mounted on a UAZ-69 truck or carried on the rear decking of an OT-62 APC. The muzzle end is provided with a bar to assist in manhandling the weapon.

The M59A is the only one in the Warsaw Pact Forces that uses a 12·7mm spotting rifle which is mounted over the barrel. The M59A is a slightly modified M59.

Employment
Czechoslovakia, Egypt.

Quad 12·7mm Anti-Aircraft Heavy MG M53

Czechoslovakia

Calibre: 12·7mm
Weight Firing: 628kg
Weight Travelling: 640kg
Length Travelling: 2900mm
Width Travelling: 1600mm
Height Travelling: 1780mm
Track: 1500mm
Barrel Length: 1069mm
Length of MG with Muzzle Brake: 1588mm
Elevation: -7° $+90^{\circ}$
Traverse: 360°
Maximum Range Horizontal: 6500m
Maximum Range Vertical: 5500m
Maximum Effective Vertical Range: 1000m
Rate of Fire (per Barrel): 550–600rpm (cyclic)
80rpm (practical)
Type of Feed: Belt, 50 rounds in each drum
Ammunition: Projectile wt 49·5 grammes, m/v 840 m/s (API round)
Armour Penetration: 20mm at 500m (API round)
Crew: 6
Towing Vehicle: TATRA-805, 4×4, Truck

This was built in Czechoslovakia and uses four Russian-designed M-1938/46 DShK heavy machine guns. It is used in a similar role as the Soviet ZPU-2 and ZPU-4 weapons.

It is recognisable by its four large drum magazines, muzzle brakes on the barrels, two-wheeled mount with levelling jacks at the front and rear. It is no longer used in Czechoslovakia.

Employment
Cuba and reported North Vietnam.

Quad 12·7mm Anti-Aircraft heavy MG M53 in firing position

Twin 30mm Anti-Aircraft Gun M53
Czechoslovakia

Calibre: 30mm
Weight in Firing Position: 2000kg
Weight in Travelling Position: 2100kg
Length Travelling: 7587mm
Width Travelling: 1753mm
Height Travelling: 1753mm
Barrel Length: 2429mm (including muzzle brake)
Elevation: $-10° +90°$
Traverse: 360°
Maximum Range Horizontal: 10,000m
Maximum Range Vertical: 7000m
Effective Vertical Range: 2000m
Rate of Fire per Barrel: 450–500rpm (cyclic)
100rpm (practical)
Feed: 10 round clips fed horizontally
Ammunition: Projectile wt HE 0·45kg, m/v 1000 m/s
Projectile wt API 0·45kg, m/v 1000 m/s
Complete rounds weight=0·91kg
Armour Penetration: 60mm at 500m
Crew: 4
Towing Vehicle: PRAGA-V3S, 6×6, Truck

Twin 30mm Anti-Aircraft Gun M53, note the screw type firing jacks under the carriage

This weapon was first seen in 1958 and is used in a similar role as the Russian 23mm guns. The twin guns have hydraulic elevation and traverse and have quick change barrels. Mounted on a four-wheeled carriage, a stowage box is on the right hand side of the mount. These weapons with a vertical feed and 50 round magazines are mounted on a PRAGA-V3S 6×6 (Armoured) when they are known as the M53/59, the vehicle being called PRAGA V3-ST.

Employment
Czechoslovakia.

57mm Anti-Aircraft Gun
Czechoslovakia

This was built in Czechoslovakia in very small numbers and was replaced by the Russian 57mm S-60 weapon. It is heavier than the latter and has a similar rate of fire but a lower muzzle velocity.

Employment
Cuba, Guinea, Mali.

57mm Anti-Aircraft Gun in firing position, note that the left of the shield folds down and that the firing jack between the wheels is in position

122mm Tampella Field Gun m/60 Finland

Calibre: 122mm
Weight: 9500kg
Elevation: -5° $+50^\circ$
Traverse: 90° (total)
360° (with additional equipment)
Range: 25,000m
Ammunition: HE shell wt 25kg, m/v 950 m/s
Towing Vehicle: Russian AT-S Tractor
Finnish Sisu KB-46, 6×6

The 122mm Tampella Field Gun m/60 is of Finnish design and construction. It is very similar in appearance to the Swedish 155mm 'F' Howitzer and the French 155mm M-1950 Howitzer.

The weapon is mounted on a heavy duty split trail, each trail has two wheels, when in the firing position a jack is let down between the two sets of wheels. For travelling, the barrel is rotated through 180°. The barrel is fitted with a muzzle brake. No shield is fitted, the recoil equipment is above and below the barrel and the weapon has hydraulic elevation. The m/60 is also used for coast defence.

Employment
Finland.

105mm Light Field Howitzer m/61 Finland

Calibre: 105mm
Weight: 1800kg
Elevation: $+6^\circ$ $+45^\circ$
Traverse: 53° (total)
Range: 13,400m
Rate of Fire: 7rpm
Ammunition: HE wt 14·9kg, m/v 600 m/s
Towing Vehicle: Sisu KB-45 (4×4) (Finnish)
Sisu KB-46 (6×6) (Finnish)

This weapon is also known as the m61-37 and is a pre-World War II Russian weapon fitted with a new Finnish barrel and cradle. The carriage has a split trail, each trail being provided with a lifting handle, a small shield is provided. The barrel has a muzzle brake fitted.

Other Artillery Weapons Used by Finland

The Finnish Army has its own designations for foreign weapons used by its Army:

m/02
Is a modified 76mm Russian gun.

m/34
Is a 152mm Russian gun, reported to be M-1910/34.

m/36
Is a 76mm Russian gun, reported to be the M-1902/30.

m/37
Is a 152mm Russian gun.

m/38
Is a 152mm Howitzer of Finnish construction with a range of 12,000m, total weight is 4200kg.

m/38
Is the 122mm Russian M-1938 Howitzer.

m/40
Is the German World War II 150mm Field Howitzer Model 40.

m/41
Is the Russian 105mm gun modified.

m/54
Is the Soviet 130mm Field Gun M-46.

ANTI-AIRCRAFT WEAPONS

m/37
Is the German World War II 88mm FlAK 36/37.

m/40
Is the German World War II 20 mm FlAK 38.

TOP
122mm Tampella Field Gun m/60 in the firing position

ABOVE
105mm Light Field Gun m/61

95mm Recoilless Rifle m/58 — Finland

Calibre: 95mm
Weight: 140kg
Barrel Length: 3200mm
Range: 700m (effective)
Rate of Fire: 6–8rpm
Ammunition: Wt 10·1kg, m/v 615 m/s
Armour Penetration: 300mm
Crew: 3

This weapon was designed and built in Finland and is mounted on two small wheels with a stabilising leg at the front and rear. To assist in moving the weapon there is a carrying handle either side, just in front of the breech, and another one either side of the barrel towards the front. The sights are on the left side of the barrel.

Employment
Finland.

81mm Mortar m/38 — Finland

Calibre: 81mm
Weight: 60kg
Range: 3000m
Rate of Fire: 20rpm
Ammunition: 3·5kg
Crew: 3–4

This is reported to be based on the German World War II 80mm Mortar Gr.W.34, although it is much different to look at. The Finnish Mortar has a circular baseplate, the bipod has a chain connecting the legs, on the left leg is a handle to alter the traverse of the Mortar, the handle to alter the elevation is at the top of the tripod. The sights are on the left side of the barrel.

Employment
Finland.

120mm Tampella Mortar — Finland

The m/41C (Swedish designation) is used by Sweden, for details refer to the Swedish section. A more recent version is reported to be used by Israel. West Germany has a number of her M-113 Armoured Personnel Carriers fitted with the 120mm Tampella Mortar, these have a range of 400–6200m.

160mm Tampella Mortar — Finland

This is used by the Finnish Army and has a weight in the firing position of 1700kg and fires a 40kg bomb to a range of 9500/13,800m. Rate of Fire is 3–4rpm. It is also built in Israel in a much modified form.

TOP RIGHT
95mm Recoilless Rifle m/58 in firing position

RIGHT
81mm Mortar m/38 and crew

105mm Light Gun

France

Calibre: 105mm
Weight: 1200kg
Length Travelling: 5120mm
Width Travelling: 1770mm
Height Travelling: 1180mm
Elevation: -5° $+70^\circ$
Traverse: 45°
Range: 15,000m with French hollow base shell Mk 63
11,600m with American HEM 1 shell

This new light gun was shown by the DTAT at the 1971 French Satory display. It has been designed for three roles: direct and indirect support, anti-tank firing (with the OCC-105-F1 non-rotating hollow charge round which is effective at 850m), and for deployment by air, either slung under a helicopter or air-dropped.

The gun is mounted on a two-wheeled carriage and can be towed in two positions either folded (travel order) or unfolded (when in combat area). Maximum towing speed is 90 km/h.

Stated times for setting into battery is less than 60 seconds and time for coming out of battery less than 30 seconds.

It is recognisable by its barrel with a double baffle muzzle brake (it has a 30 calibre barrel), no shield and split box type trails.

NOTE: It has also been seen fitted with a shield.

Employment
Trials.

Weapon in firing position

75mm Field Gun M-1897 — France

Calibre: 75mm
Weight Travelling: 1139kg
Length Travelling: 5577mm
Width Travelling: 2011mm
Height Travelling: 1402mm
Elevation: −11° +18°
Traverse: 6° (total)
Range: 8550m (m/v is 624 m/s)
Rate of Fire: 12rpm

This is the famous 'French 75' which was used in both World Wars. The early models had spoked wheels with metal rims but later they were fitted with new axles and rubber tyres. The barrel has no muzzle brake. It was also built in the United States and Mexico.

Employment
Cameroon, Cambodia, Bulgaria, Greece, Morocco, Laos, Mexico, Upper Volta.

105mm Howitzer M-1950 — France

Calibre: 105mm
Weight Travelling: 2922kg
Length Travelling: 6004mm
Width Travelling: 1981mm
Height Travelling: 1706mm
Elevation: −5° +70°
Traverse: 360°
Range: 14,500m (m/v 580 m/s)
Rate of Fire: 6rpm

This was developed after the end of World War II and is very similar in some respects to the German 105mm Feldhaubitze 43 (L/28). It is recognisable by its curved shield, single-tyred wheel each side, three trails (one behind the breech and two in front of the shield), and torsion bar suspension. The barrel has a double baffle muzzle brake.

Employment
Cameroon, France, Guinea, Israel, Ivory Coast, Morocco.

105mm Howitzer M-1950

155mm Field Howitzer M-1950 — France

For full details of this, see the Swedish section, as this gun is built in Sweden under the designation Model 'F'. It is also used by Israel, Lebanon and Tunisia, in addition to France. It is also mounted on an AMX chassis, it is then known as the 155mm Self-Propelled Howitzer. When in the self-propelled role, it has a different muzzle device.

20mm Gun, M-621 on commando mount

20mm Gun, M-621

France

Calibre: 20mm
Weight: 140kg (excluding gun)
Length Travelling: 2350mm
Track: 1410mm
Elevation: -30° $+26^\circ$
Traverse: 360° (zero elevation)
95° (negative elevation)
70° (positive elevation)
Rate of Fire: 300 or 740rpm
Ammunition Types: HE, HE Tracer, Incendiary
Perforating shell
m/v 1000+ m/s Cartridge wt 250gr
Armour Penetration: 25mm at 1000m
Crew: 2
Towing Vehicle: 4×4 Jeep
NOTE: The above data relates to the M-621 mounted on the Commando mount.

When on the Commando mount the weapon can be towed by a vehicle, carried by a helicopter, and can also be broken down into smaller loads. The weapon is equipped with a collimator.

Basic data of the gun is: weight gun and cradle 58kg, length gun and cradle 2207mm width gun and cradle 202mm, height gun and cradle 245mm. Can fire either single rounds or bursts.

Other types of mount to which the M-621 can be fitted include the 13 A helicopter mount Mk 70, T 20 13 cupola for the AML M3 APC, mounted on Alouette 111 mount, HE 60-20 turret on AML-60 Armoured Car, S-530 twin gun turret on AML-60 chassis, truck mount, Jeep mount, coastal boat mount and pontoon boat mount.

A more recent development of the M-621 is the M-693. This is being fitted to the AMX-10P Infantry Combat Vehicle. This has a rate of fire of 700/780rpm, weight of weapon and cradle is 80kg, ammunition length is 207mm, muzzle velocity 1300 m/s.

Employment
In production.

120mm Brandt Mortars — France

	M-120 M 65	M-120-RT-61	AM 50	MO-12-60
Weight inc Carriage:	144kg	580kg	402kg	—
Weight Firing:	104kg	580kg	242kg	92kg
Weight Tube and Breech:	44kg	114kg	76kg	34kg
Weight of Mount:	24kg	257kg	86kg	25kg
Weight of Baseplate:	36kg	190kg	86kg	33kg
Weight of Carriage:	40kg	—	154kg	—
Barrel Length inc Breech:	1640mm	2080mm	1746mm	1632mm
Elevation:	+40° +85°	+28° +85°	+45° +80°	+40° +85°
Traverse—mils:	300	250	300	300
Maximum ROF, rpm:	12	10	12	15
Average ROF, rpm:	8	6	8	8
Time to Employ:	—	3 minutes	—	—

The above mortars are all designed and manufactured by the French Company Thomson Houston (Hotchkiss Brandt).

M-120 M65

This is transported on a two-wheeled carriage, this has small pneumatic wheels and elastic suspension, the carriage is removed for firing. It is towed muzzle first by a light vehicle or two men. For transportation over rough ground it can be broken down into three man-pack loads. It is recognisable by its triangular baseplate and small wheeled carriage. The mortar can fire the following bombs:

PEPA/LP

Bomb weight 13·42kg, maximum range with charge 6 is 9000m, minimum range 1200m. PEPA stands for Projectile Empenne à Propulsion Additionelle (Additional Propulsion).

M 44 HE

Bomb weight 13kg, range 6650m, minimum range 500m. Seven charges.

Illuminating Mk 62

Weight 13·60kg, range 5000m, minimum range 900m.

Smoke

Weight 13kg, maximum range 6650m, minimum range 500m. Also marking bombs and practice bombs, characteristics similar to the M 44.

120mm Brandt Mortar MO-120 M 65

M-120-RT-61

This is a rifled mortar and when in firing position its wheels are retained in position. The weapon consists of three parts: barrel with breech and towing ring, mount (cradle and undercarriage), and baseplate (triangular). It fires the following bombs:

PRPA

Total weight 18·70kg, maximum range 13,050m. This has additional propulsion. PRPA stands for Projectile Rayé à Propulsion Additionelle.

RP-14

Weight 18·70kg, maximum range 8135m, minimum 1200m. The weapon can also fire the M 44, PEPA, PEPA/LP, Smoke and Illuminating bombs.

120mm Brandt Rifled Mortar MO-120-RT-61

PRECLAIR (Projectile rayé éclairant)

This is an illuminating bomb and its shell weight is 15·45kg. Burning time is 60 seconds, and its candle power for that burning time is of 1,050,000 candles or 40 seconds with 1,600,000 candles. Its diameter of illumination is of 1000m. The minimum range is 300m and maximum range 8000m.

120mm Mortar AM 50 (MO-120 AM 50)

This weapon can be towed by its muzzle and can be fired with or without its wheels. It is recognisable by its two-wheeled carriage and triangular baseplate, when in the travelling position the barrel is in a horizontal position. Fires the PEPA/LP, M 44, Smoke bomb Mk 62 and Illuminating Mk 62.

120mm Light Mortar MO-120-60

This weapon breaks down into three loads for easy transportation, i.e., tube and breech, bipod, and baseplate. There is also a special trailer to carry the mortar and 12 rounds of ammunition. It is recognisable by its star-shaped baseplate, bipod with chains connecting the two legs, twin short recoil cylinders under the barrel and the anti-cant device attached

120mm Brandt Mortar AM 50 (MO-120 AM 50)

to the left leg. Fires the M 44 bomb but only up to charge 4, PEPA/LP bomb but only to charge 4, Illuminating bomb Mk 62 but to charge 4 only, PEPA bomb (minimum 600m, maximum 6550m range), also smoke, practice, target indicating, and coloured.

Employment
Used by many countries and also built under licence overseas.

These mortars are also mounted in various vehicles, for example the French AMX VCI has been fitted with a 120mm mortar.

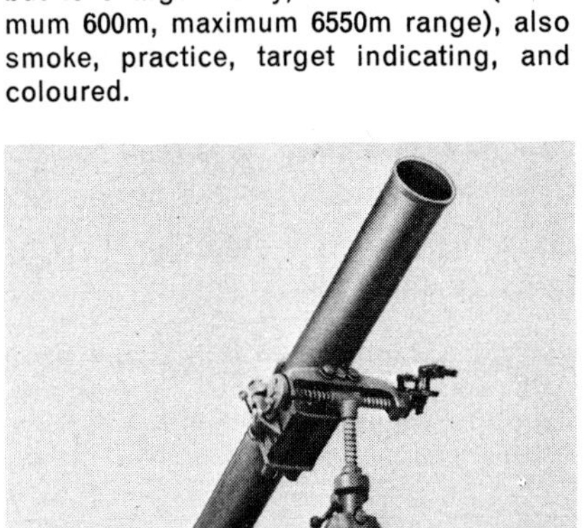

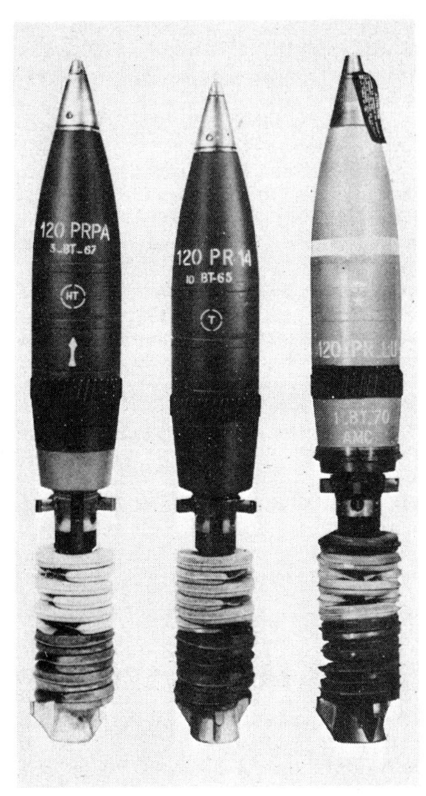

ABOVE
120mm Brandt Light Mortar MO-120-60

RIGHT
120mm rifled projectiles for the 120mm Brandt Mortars

BELOW
Projectiles for the 120mm Brandt Mortars

81mm Light Mortar — France

Calibre: 81mm
Weight Total: 39·40kg (short barrel) 41·50kg (long barrel)
Barrel Weight: 12·40kg (short barrel) 14·50kg (long barrel)
Weight of Baseplate: 14·80kg
Weight of Mount: 12·20kg
Barrel Length: 1150mm (short barrel) 1450mm (long barrel)
Elevation: +30° +85°

This mortar is produced by Thomson Houston (Hotchkiss Brandt). There are two models—the short version (MO-81-61-C) or the long version (MO-81-61-L). The weapon can be quickly broken down into three loads for easy transportation, i.e. barrel, baseplate and mount. The weapon is recognisable by its light, triangular-shaped baseplate and bipod with chains connecting the legs. The 81mm mortar can also be mounted on vehicles, i.e. the AMX VTT APC.
The weapon can fire the following types of bombs:

HE Bomb Type M 57 D. Weight 3·30kg, maximum range 4100m, minimum range 120m. Filling is tolite, this bomb being for the short version.

HE Bomb Type Mk 61. Weight 4·325kg, maximum range 5000m, minimum range 75m, designed for the long-barrelled version although it can also be used for the shorter version. Tolite filling.

Coloured HE Bombs M 57 D or Mk 61. Colours are red, green or yellow.

Smoke Bombs M 57 D or Mk 61. Data similar to HE bombs.

Practice Bombs Type M 57 D or Mk 61. Data similar to HE bombs. Filling is a mixture of sulphur and naphthalene with dummy fuses.

Illuminating Bombs.
Mk 62, weight 3·150kg, range 500–3400m, 250,000 candle power, minimum burning period 40 seconds.
Mk 68, weight 3·50kg, range 600–4100m, 400,000 candle power, minimum burning period 40 seconds.
Both have a filling based on magnesium and a lighting radius of 250m.

Employment
Short-barrelled version used by the French Army.

81mm Model 1944 Light Mortar — France

This has a calibre of 81mm, barrel length of 1150mm, weight in firing position of 59·7kg, and fires a bomb of 3·30kg to a range of some 3·15km.

Employment
Cambodia, Chad, Dahomey, Egypt, Gabon, Ivory Coast, Laos, North Vietnam, South Vietnam, Upper Volta.

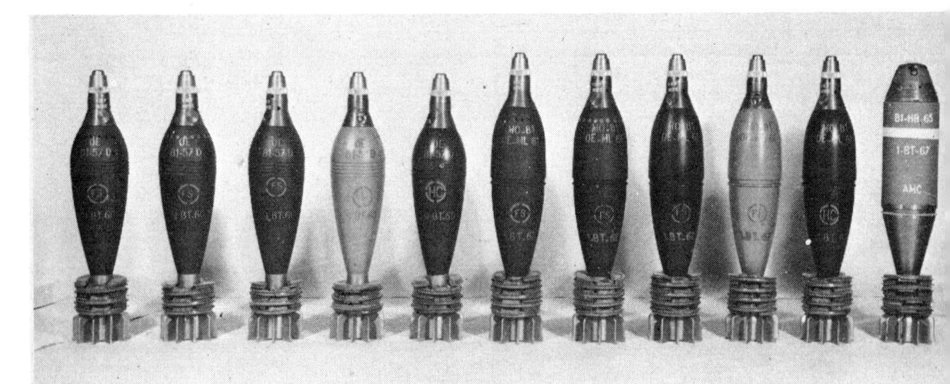

LEFT
Brandt 81mm Mortar Bombs

ABOVE
81mm Light Mortar Long Barrelled Version

RIGHT
81mm Light Mortar Short Barrelled Version

LFH-18 in firing position

105mm Light Field Howitzer 18 (LFH-18)

Germany

Calibre: 104·9mm
Weight Travelling: 1985kg
Length Firing: 5994mm
Length Travelling: 5588mm
Width Firing: 3580mm
Width Travelling: 2010mm
Height Travelling: 1800mm
Barrel Length: 2706mm (25·77 cal)
Elevation: $-5° +42°$
Traverse: 56° (total)
Maximum Range: 10,600m
Rate of Fire: 6–8rpm
Ammunition: HE, AP, Smoke etc
HE projectile wt 14·8kg, m/v 670 m/s
Crew: 6–10
Towing Vehicle: Truck

The 105mm LFH-18 was first built in 1935 by Rheinmetall, it was the standard German Divisional Field Howitzer. The recoil and recuperator are of the hydro-pneumatic type. The trails are of the split type and the spades, when in the travelling position, are folded forward onto the tops of each trail. The weapon is provided with a small shield. The carriage has large wheels built of light alloys, these are fitted with solid rubber tyres.

The 105mm LFH-18 was modified as the LFH-18 (M), also known as the LFH-18 (40). This is recognisable as the barrel has a muzzle brake, the wheels are spoked and have solid tyres. It has a range of 12,200m.

Employment
Argentina, Austria (LFH-18 (M)), Chile, Portugal, Sweden (known as Haub m/39), Yugoslavia (both LFH-18 and LFH-18 (M)).

90mm Self-Propelled Anti-Tank Gun
Federal German Republic

Calibre: 90mm
Weight: 5000kg
Length Firing: 7300mm
Width Firing: 2500mm
Height Firing: 1400mm
Ground Clearance: 380mm
Fording Height: 990mm
Elevation: $-8° +25°$
Traverse: 50° (total)
Crew: 3
Speed: 20kmph
Gradient: 50%

This was developed by Rheinmetall in the 1960s and only prototypes were built.

Basically it consists of a chassis on which is the engine, three seats for the crew, ammunition, steering gear and braking system. Onto this is fitted the gun and shield. The weapon has two large wheels at the front and two smaller ones at the rear. It can be steered in two ways—first with the twin rear wheels via the steering wheel and second by the steering levers. In addition the weapon can be towed by a truck. On the shield are eight smoke dischargers, conventional as well as infra-red sights are fitted.

When required for firing the chassis is lowered at the rear, rear wheels retract to the rear, recoil spades (one either side, both hydraulic) are let down and the barrel unlocked and goes forward. To take out of action the spades are retracted, the rear wheels fully in position and the barrel retracted and locked.

The gun uses the same ammunition as the German 90mm Kanone and M-48 tank.

ABOVE
Gun in firing position, note position of the wheels

BELOW
Gun with barrel retracted

Germany (World War II)

20mm Flak 38 (Anti-Aircraft Gun)

This consisted of four 20mm guns mounted on a quadruple mounting. They were on a two-wheeled trailer which was provided with levelling jacks, it was introduced into service in 1941. Basic data is weight travelling: 2217kg, weight firing: 1450kg, length: 4300m, width: 2400mm, height: 2135mm. The guns have a traverse of 360° and an elevation between −10° and +100°. Rate of fire is 800rpm (practical), a total of 320 ready rounds are carried. The maximum horizontal range is 4790m and maximum vertical range is 3600m. The HE round has a muzzle velocity of 628 m/s and the AP round a m/v of 800 m/s. The full designation of this weapon is Flakvierling 38. The Flak 38 is reported to be used in Finland under the designation m/40. There is also a single version of this weapon which may be in service with such countries as Yugoslavia.

50mm Pak 38 (Anti-Tank Gun)

This was introduced in 1941 and the barrel has a double baffle muzzle brake, the breechblock is of the horizontal sliding type. It is mounted on a welded carriage with split trails, when the trails are opened out the torsion bar suspension is automatically locked. It has metal disc wheels and solid tyres. The shield has spaced armour and its sides slope to the rear. Basic data is calibre: 50mm, weight: 914kg, total traverse: 80°, elevation: −4° to +22°. AP round had a m/v of 792 m/s and the HE round a m/v of 548 m/s. A third wheel could be added to the rear of the tubular trails. It is reported that this weapon is still used in Spain and Yugoslavia.

75mm Pak 40 (Anti-Tank Gun)

This has a carriage with tubular split trails with a spade at the end of each trail. The wheels are of light alloy and are fitted with solid rubber tyres. The shield has spaced armour and the sides slope to the rear, the axle is protected by a hinged piece of armour. The barrel has a double baffle muzzle brake and the breechblock is of the horizontal sliding type. Weight is 1500kg, traverse (total) 65° and an elevation of −5° to +22°. It is reported to be still used in Yugoslavia.

88mm Flak 18, 36 and 37 (Multi-Role Gun)

This is the world-famous 88 gun which was developed in many versions. The 18, 36 and 37 were all mounted on carriages with four dual wheels and outriggers. Brief data is weight travelling: 7800kg, weight firing: 5600kg, length travelling: 7700mm, height travelling: 2590mm, height firing: 1600mm, width travelling: 2400mm, traverse: 360°, elevation: −3° to +85°. Rate of fire 15–20rpm. Maximum horizontal range: 14,800m, vertical range: 10,600m. Ammunition types included HE (m/v 820 m/s) and at least three types of AP (m/v 800 m/s). Still reported to be used in Finland, Spain, Yugoslavia and Czechoslovakia (reserve); m/37 in Finland.

105mm Flak 38 (Anti-Aircraft Gun)

This is reported to be held in reserve in small numbers in Czechoslovakia. Basic data was calibre: 105mm, weight travelling: 14,600kg, weight firing: 10,200kg, length travelling: 8400mm, height travelling: 2895mm, height firing: 1800mm, width travelling: 2440mm, traverse: 360°, elevation: −3° to +85°. Rate of fire: 12–15rpm. Maximum horizontal range: 17,710m, maximum vertical range: 12,730m. Ammunition: HE (m/v 880 m/s) and AP.

15cm s.F.H.18 Field Howitzer

This weapon has a hydropneumatic recoil mechanism and a horizontal sliding breechblock mechanism. When travelling a two-wheeled limber is attached to the rear and the barrel is moved out of battery. All of the wheels have solid rubber tyres. It has a weight of about 5600kg and a range of 13,300m, and can fire a HE or AP round.

The weapon is still used by Finland (known as the m/40 and much modified), Yugoslavia, Portugal and Czechoslovakia. The Czechs have fitted it with a new barrel and call it the 152mm Howitzer M 18/46, for full details of this see the Czechoslovakian section. Actual calibre of the s.F.H.18 was 149mm.

20mm Cannon, Mk 20, Rh.202 (Twin Mount)

Federal German Republic

Calibre: 20mm
Weight Travelling: 2100kg (including ammunition)
Weight Firing: 1450kg (including ammunition)
Length Travelling: 4500mm
Width Travelling: 2370mm
Height Travelling: 1900mm
Track: 2080mm
Barrel Length: 1840mm (excluding muzzle brake)
Elevation: $-5°$ $+83°$
Traverse: 360°
Tactical Range of Engagement: 1000–2000m
Maximum Range: 7000m
Rate of Fire: 800–1000rpm per barrel (cyclic)
Ammunition: HEI-T, m/v 1050 m/s
API, m/v 1100 m/s
TP-T, m/v 1050 m/s
TP, m/v 1050 m/s
Armour Penetration: AP round at 1000m
NATO angle 0° 32mm Armour Plate
NATO angle 30° 24mm Armour Plate
NATO angle 60° 8mm Armour Plate
Ammunition Supply: 270 rounds per barrel
Crew: 1 man on gun
2 men replenishing ammunition
Towing Vehicle: Mercedes-Benz Unimog, 4×4, Truck

This anti-aircraft system has been designed and manufactured by Rheinmetall for the German Army and Air Force. The weapon system consists of 2×20mm guns with their ammunition supply mounted on a cradle. This is mounted on the upper chassis which is in turn mounted on a lower chassis. This lower chassis has three outriggers, of these, two are adjustable for height and they support the

Twin mount in firing position

weapon when firing. The lower chassis is mounted on a two-wheeled carriage for transportation.

The upper chassis also carries the gunner's seat, laying mechanism (including an air-cooled single cylinder Wankel petrol engine, 8hp at 4500rpm, hydrostatic drives for elevation and traverse, hydraulic dampers, oil tank etc, should this fail the guns can be laid manually), and the fire control system. The latter consists of a lead computer, target tracking device, the optical sight and the drive for the lead marks in the sight. It also has a 'taboo' feature so that friendly targets will not be fired upon.

The weapon has hydraulic elevation and traverse, traverse being 100° a second and elevation 55° a second. Can also be traversed and elevated manually. There is a selector switch for single shots or bursts, on both guns or each gun. Electrical or mechanical firing. The A/A sight is magnification 1:1, field of view azimuth 45°, elevation 40°, for ground role a telescopic sight magnification 4, field of view 12° with a range scale up to 4000m.

The Mk 20, Rh. 202 guns are also used in various other roles including: shipboard single or double mounts, turret mounted (TS-10) on M-113 APC, A/A armament for the now defunct MBT-70, turret mounted (TS-7) in the new Radspahpanzer, and turret mounted on the Marder APC. Can also be mounted on a single mount.

Employment

Federal German Army and Air Force.

Single mount in firing position

Twin mount being towed

A 5·5in Gun in action during World War II

5·5in Gun

Great Britain

Calibre: 140mm
Weight: 5850kg
Length: 7518mm
Width: 2540mm
Height: 2616mm
Barrel Length: 4343mm
Elevation: −5° +45°
Traverse: 30° left and right
Range: 14,800m with 45·35kg shell
16,460m with 36·28kg shell
Rate of Fire: 2rpm
Ammunition: HE, Smoke, Illuminating (separate loading)
Crew: 10
Time into Action: 10 minutes
Towing Vehicle: AEC Matador, 4×4, Truck
Leyland, 6×6, FV 1103 Truck

Official designation is Ordnance, BL, 5·5in. The requirement for a 5·5in gun/Howitzer was made in 1939. The 5·5in used the same carriage as the obsolete 4·5in gun, the carriage being designed originally for a 4·5in and 5in weapon. The weapon was approved for production in August 1939 and saw action from 1941 in Africa. It was the standard weapon of the medium regiments of the Royal Artillery until the late 1960s, it is, however, still used by the Royal Artillery for training and reserve roles.

The weapon is recognisable by its split trails, spades being carried on the top of each trail whilst travelling, large single tyres, barrel has no muzzle brake. Two types of balancing gears were fitted, hydropneumatic or spring, the breech-block is of the interrupted screw thread type.

Employment
India, Malaysia, New Zealand, Oman, Pakistan, South Africa and Burma.

Recoilless Rifles

During World War II Britain developed a whole range of recoilless weapons. None of these, however, entered service. The first weapon to enter service was the 120mm L2 BAT (Battalion Anti-Tank) in the early 1950s. This was mounted on a two-wheeled carriage and towed by its muzzle. As far as it is known none of these remain in service today.

120mm L4 Mobat

This followed the L2 and was introduced into service in the 1950s and is still used in some numbers by the British Army (both in front line and reserve units). It is mounted on a two-wheeled carriage and towed by a 4×4 Land Rover by its muzzle. There is an optical sight on the right hand side and a 7·62mm spotting weapon (a modified Bren) on the left hand side. The breech Venturi drops for loading and the round of ammunition is slid into the breech using the top of the venturi. Basic data of the Mobat is:

Calibre: 120mm
Weight: 746kg
Length: 3450mm
Width: 1525mm
Height: 1190mm
Range: 800m
Crew: 3

120mm L6 Wombat

This was developed from the Mobat and is lighter and has a longer range, it was introduced into service in the early 1960s. The Wombat has replaced the Mobat in many units. It is mounted on a small two-wheeled mount, it is not designed to be towed. It is carried in the rear of a long wheel base Land Rover, and may be fired from the vehicle in an emergency, the vehicle carries six rounds of ready use ammunition. Two small ramps can be placed at the rear of the Land Rover, and with the aid of a winch on the vehicle the Wombat can be quickly unloaded. It can also be mounted and fired from the FV 432 APC. In recent months the Wombat has been fitted to the new lightweight Land Rover and to the Bv.202 tracked vehicle. A recent development of the Wombat is the L8 Conbat.

The Wombat is fitted with an American ·50 (12·7mm) spotting rifle over the barrel and the breech opens to the right. The ammunition is the same as that used in the Mobat and the HESH round weighs 27·2kg, has a muzzle velocity of 460 m/s and the projectile itself weighs about 12·7kg. Basic data of the Wombat is:

Calibre: 120mm
Weight: 294kg
Length: 3860mm
Width: 863mm
Height: 1092mm

120mm Mobat showing the spotting rifle

Range: 750m (moving targets)
1000m (static targets)
Rate of Fire: 4rpm
Crew: 3

Employment

L-4 Mobat	Great Britain, Jordan, Kenya, New Zealand, Malaysia
L-6 Wombat	Great Britain and Australia

BELOW
120mm Wombat being unloaded from a Land Rover, the man on the right is operating the winch

FOOT
120mm Wombat in the ready to fire position on a Land Rover, the 12·7mm spotting rifle can be clearly seen

105mm Light Gun

Great Britain

Calibre: 105mm
Weight Travelling and Firing: 1768kg
Length Firing: 7010mm
Length Folded for Travelling: 4080mm
Length Travelling (Gun Forward): 8800mm
Width: 1778mm
Height Travelling (Folded): 1229mm
Height Travelling (Gun Forward): 2133mm
Track: 1410mm
Elevation: $-5\frac{1}{2}^\circ$ $+70^\circ$
Traverse: $5\frac{1}{2}^\circ$ left and right
360° on platform
Range: 2500m minimum
15,000m maximum
17,500m with super-charge at present under development
Rate of Fire: 3rpm sustained
6rpm maximum
Ammunition: HE (wt 15·88kg)
White phosphorus
Base Ejection Smoke
HESH, HESH Practice
Illuminating
Target Marker
Anti-Personnel (canister)*
Crew: 6
Towing Vehice: 1 tonne, 4×4, Land Rover (also $\frac{1}{4}$ and $\frac{3}{4}$ ton Land Rovers)
Bv 202 Tracked Vehicle
Airtransportable: SA 330 Puma (one load)
Sea King (one load)
Wessex Mk 2 or Mk 5 (two loads)
* Under development.

The current British Army Light Gun is the Italian 105mm Pack Howitzer. In 1965 it was decided to develop a new 105mm gun to replace the Italian weapon. Prototypes were built and tested and the weapon was accepted for British Army use in 1971 and it is expected that it will enter service in 1974.

The 105mm gun has two towing attitudes. First is the normal folded position. In this the equipment is jacked up (the jack being stowed on the trail), the quick release, right hand wheel and traverse gear pin removed, and the gun swung through 180°. The barrel is then clamped to the trail and the wheel replaced. This takes less than one minute. To convert from the firing position to the towing attitude the two links are removed, the equipment pushed off the platform, this is then secured to the inside of the trail and the links refitted for travelling. This takes

105mm Light Gun on its turntable

less than one minute. The 105mm gun can be split into two for helicopter loads and reassembled with simple tools in less than 30 minutes.

The barrel has a double baffle muzzle brake and a hand-operated vertical sliding breechblock. This is operated by a lever mounted on the top. The hydropneumatic recoil system has a separate recuperator. Electric firing gear is fitted. Direct and indirect sights are provided. The wheels have hydraulic brakes for towing, these can be operated by a lever at the end of the trail whilst firing. The carriage is fitted with trailing arm suspension with shock absorbers. The platform enables the gun to be quickly traversed through 360°, it is connected to the underside of the chassis by three wire stays.

A special barrel has been developed to use up the remaining stocks of US M1 ammunition. This takes two hours to change and will be used for training.

Recognition features are its bow-shaped tubular trail with spade, circular traversing platform, no shield, double baffle muzzle brake, balancing springs either side of the rear of the gun, when travelling barrel is to the rear.

105mm Light Gun being fired

25 Pounder Field Gun — Great Britain

Calibre: 88mm
Weight: 1800kg (including firing platform)
Length: 7924mm
Width: 2120mm
Height: 1651mm (travelling)
1695mm (firing)
Ground Clearance: 342mm
Barrel Length: 2295mm
Elevation: -5° $+40^{\circ}$
Traverse: 4° left and right
360° on turntable
Range: 12,250m
Rate of Fire: 5rpm
Ammunition: HE (wt 11·34kg), AP, Smoke
Crew: 6
Towing Vehicle: Ford 3 ton, 4×4, Truck
Bedford 3 ton, 4×4, Truck
Morris Quad, 4×4, Truck

In the late 1920s and early 1930s studies were conducted to find a replacement for the 18 Pounder Gun and 4·5in Howitzer used by the Royal Artillery. The prototype 25 Pounder was completed in 1937, this had split trails, but the weapon was found to be too heavy. Instead the carriage of an abandoned 4·1in Howitzer was used. The 25 Pounder was approved for production in December 1938.

Before sufficient 25 Pounders became available 18 Pounders were relined and mounted on their original carriages, these were known as 18/25 Pounders or 25 Pounders Mk 1 and had a range of some 11,700m. The next models were the first complete 25 Pounder and the 25 Pounder Mk 2, the barrel had no muzzle brake and the trail was of the box type. The next model was the Mk 3, this had a muzzle brake. Over 12,000 25 Pounders were built during World War II and the weapon did not pass out of front line British Army use until 1967, it is, however, still used by the British Army for training purposes.

Other versions of the 25 Pounder included:
25 Pounder mounted on a Valentine chassis, this was called the Bishop and saw action in North Africa and Italy.
25 Pounder mounted on a Canadian Ram chassis and called the Sexton, saw service during and after World War II.
25 Pounder on a Loyd carrier chassis, trials only.
25 Pounder on a Centurion chassis (shortened) (FV 3802), prototypes built post-war for trials only.
25 Pounder on FV 300 chassis, FV 304, prototype only, post-war.

There was also the 'Baby 25 Pounder', this had a cut-down barrel, shorter trail with a caster, no shield and a flash hider was fitted on the barrel, this was developed for jungle warfare and had a shorter range. Another version was the Indian 25 Pounder (Baby) and a Indian/Canadian airborne version.

During World War II the 25 Pounder was normally towed by a Morris Quad 4×4 Truck, this also towed the 'limber' (Artillery Trailer No 27), the firing platform could also be carried on the limber.

25 Pounder Mk 2 Field Gun. Note the turntable under the trail.

Employment
Australia, Bangla Desh, Burma, Cyprus, Egypt, Eire, India, Indonesia, Israel, Jordan, New Zealand, Netherlands, Nigeria, Malaysia, Pakistan, Portugal, Rhodesia, South Africa, South Yemen, Singapore, Sudan, Oman, Ghana, Abu Dhabi, Ras Al Khaimah, Sharjah.

NOTE: *It is reported that South Africa has modified some of their 25 Pounders, they are reported to be called 90mm Field Gun.*

ABOVE
An Indian 25 Pounder in action in the Indo-Pakistan Campaign of late 1971

BELOW
An Indian 25 Pounder in action in the Eastern sector during the Indo-Pakistan Campaign of late 1971

6 Pounder Anti-Tank Gun — Great Britain

The design dates from 1938 and the first production gun was completed in 1941, entered service in 1942, it was also used as a tank gun. It was manufactured in the United States under the designation 57mm Gun M-1, the M-1 had hand traverse and commercial type tyres, next came the M-1A1 which had combat tyres, the M-1A2 had no handwheel traverse and was the same as the British weapon, the M-1A3 had a slightly different towing attachment. The weapon has a small shield, two wheels and a split trail. It was also mounted on an armoured truck known as the Deacon, the British also had a small SP version. Basic data is:

Calibre: 57mm
Weight Travelling: 1224kg
Length Travelling: 4724mm
Width Travelling: 1889mm
Height Travelling: 1280mm
Elevation: −5° +15°
Traverse: 90° (total)
Range: 8990m (maximum)
Rate of Fire: 20rpm
Armour Penetration: 118mm at 475m
Towing Vehicle: Jeep

Employment
M-1 series: Brazil, Thailand and Yugoslavia. 6 Pounder: Bangla Desh, Burma, Cameroon, Egypt, India, Israel, Malaysia, Pakistan and Spain.

17 Pounder Anti-Tank Gun — Great Britain

Developed in 1940/1941 and entered service in 1942, some early models were on a 25 Pounder carriage. Also mounted on the Achilles (modified M-10) and Archer for SP role, there was also the Monitor and Amazon experimental SP versions. Basic data is:

Calibre: 76·2mm
Weight Travelling: 3040kg
Length Travelling: 7540mm
Width Travelling: 2225mm
Height Travelling: 1676mm
Elevation: −6° +16½°
Traverse: 60° (total)
Range: 10,500m (maximum)
Rate of Fire: 20rpm
Armour Penetration: 222mm at 1000m
Towing Vehicle: Truck

Employment
Burma, Egypt, Israel, Pakistan, South Africa.

3·7in Anti-Aircraft Gun — Great Britain

This was developed before World War II and entered production in 1938. There were three Marks, 1, 2, 3, and four types of mount, i.e. for field or fixed roles. The weapon is recognisable by its four-wheeled carriage and its four out-riggers. When travelling these are vertical. The 3·7in could also be used in the anti-tank role. It was replaced in the British Army by the Thunderbird Missile. Basic data is:

Calibre: 94mm
Weight Travelling: 9400kg
Elevation: −5° +80°
Traverse: 360°
Maximum Range: 18,800m (horizontal)
12,000m (vertical)
9000m (vertical effective)
Rate of Fire: 8–20rpm
Ammunition: HE, AP or shrapnel, HE m/v=792 m/s
Towing Vehicle: Matador, 4×4, Truck

Employment
Reported to be used by Burma, Egypt, India and Malaysia.

TOP
British 6 Pounder Anti-Tank Gun

ABOVE
British 17 Pounder Anti-Tank Gun

BELOW
British 3·7in Anti-Aircraft Guns (World War II photograph)

3in Mortar

Great Britain

The 3in (81mm) Mortar was developed before World War II and other marks were developed during the war, for example the Mk IV with new sights and the Mk V for use in jungle warfare. The weapon was replaced in the British Army by the 81mm Mortar in the early 1960s. Basic data of the 3in Mortar is:

Calibre: 3in (81mm)
Weight (Total): 56·3kg
Elevation: +45° to +80°
Traverse: 36°
Range: 2500m
Rate of Fire: 10rpm
5rpm (sustained)
Ammunition: HE, Smoke and Chemical
Crew: 3

Employment
Bangla Desh, Burma, Ceylon, Cameroon, Egypt, India, Indonesia, Israel, Malaysia, Nigeria, Muscat and Oman, Pakistan, Saudi Arabia, South Yemen.

3in Mortar in firing position

4·2in Mortar

Great Britain

The 4·2in (106mm) Mortar was developed in 1940/1941 to fire a chemical round, the first production Mortars were completed in 1942. The Mk 1 version had a baseplate and bipod, but the Mk 2 version was fitted with an unusual two-wheeled trailer developed by Jowett Cars Limited. The Mk 2 version was used in the British Army until the 1960s, and was towed by a Land Rover or a 1 ton Truck. Mk 2 basic data is:

Calibre: 106mm
Weight of Barrel: 41·3kg
Weight of Bipod: 20·8kg
Total Weight Travelling: 550kg
Barrel Length: 1980mm
Range: 960–3750m
Ammunition: Chemical or HE (9kg)
Crew: 6

81mm Light Mortar

Canada/Great Britain

Calibre: 81mm
Weight (without Sight): 35·4kg
Barrel Weight: 12·3kg
Bipod Weight: 11·8kg
Baseplate Weight: 11·4kg
Overall Barrel Length: 1280mm
Outside Diameter of Barrel (Muzzle End): 86mm
Outside Diameter of Barrel (Breech End): 94mm
Bipod Length (Folded): 1143mm
Baseplate Diameter: 546mm
Elevation: +45° to +85°
Traverse: 100 mils left and right at 45° elevation
Crew: 2–3

The 81mm Mortar was designed in Britain and Canada some 12 years ago to replace the 3in Mortar in the British Army. Britain designed the barrel and bipod and Canada the baseplate and sight (Dial Sight C2

ABOVE
British 4·2in Mk 1 Mortars, photograph taken during World War II

BELOW
Men of the 2nd Battalion Royal Irish Rangers in action with the 81mm Mortar

81mm Mortar being fired

weight 1·25kg). The 81mm Mortar is more effective and much lighter than the 3in Mortar.

The weapon can be broken down into three parts (man packs) for easy transportation, i.e. barrel, bipod and baseplate.

The barrel is of high-strength steel with cooling fins on the lower half, the bipod is of special steel and light alloy construction. The baseplate is forged in light alloy and has four deep webs, it is circular in shape with its centre dished, in the centre is the socket for holding the barrel.

Ammunition available includes:
High Explosive (HE) L15A3, weight 4·4kg, range 200–5600m. Body is of ductile cast-iron.
Smoke (WP) L19A1, weight 4·5kg, body is of graphite cast-iron.
Practice, L22A1, range 80m.

The increments are clipped onto the tail of the bomb. There are a total 6 primary charges, minimum range 200m, maximum range 4500m. There are two new charges, 7 and 8, these increase the range to 5200m and 5600m respectively. Rate of fire is 15 rounds per minute. The 81mm Mortar is also fitted in the British FV 432 APC.

Employment
Abu Dhabi, Canada, Great Britain, Guyana, India, Kenya, Malaysia, New Zealand, Nigeria, Ras Al Khaimah, South Yemen. The 81mm Mortar is designated L1A1 in British service.

Type A32 (M) Twin 30mm Anti-Aircraft Guns

Great Britain

Calibre: 30mm
Weight: 3640kg (w/o gunner or ammunition)
Length Travelling: 7600mm (including tow bar)
Width Travelling: 2400mm
Height Travelling: 2300mm
Height Emplaced: 1700mm (wheels removed)
Wheelbase: 3700mm
Elevation: −15° +80°
Traverse: 360°
Range: 3000m (effective)
Rate of Fire: 600–650rpm per barrel (cyclic)
Ammunition: HE-P An inert practice shell
HET-P An inert shell with tracer
HEI An HE shell (HS=UAI)
HEIT An HE shell with tracer (HS=IAIT)
APICT An AP incendiary shell with hard metal core and tracer (HS=RINT)
APHEI An AP incendiary shell with self-destruction base fuse (HS=RIA)
Ammunition Supply: Type A=2 magazines of 80 rounds
Type B=2 magazines of 140 rounds+20 rounds per gun in feed chutes
Crew: 1 on gun
Towing Vehicle: 4×4 Truck, i.e. Bedford

This weapon has been developed by the British Manufacture and Research Company of Grantham, Lincolnshire. It is developed from the gyro-stabilised twin 30mm naval mounting type A32 which is fitted to a variety of naval craft.

The modifications to the naval mounting

A32(M) Guns

A32 include the addition of an armoured shield for the local gunner and provision of supplying power to the mounting in various voltages, for example AC current or 28V DC.

The guns used in this system are the Hispano-Suiza type 831 L. These are of a similar model to those of the French AMX-13 and AMX-30 anti-aircraft tanks, the British Falcon anti-aircraft vehicle and the French AML armoured car Model H 30 turret.

The director can control all functions of the gun mounting including the accurate tracking of the guns and the firing of the weapons. Local control of the gun by visual tracking is possible in the case of an emergency. Elevation velocity is 60° a second and traverse velocity is 90° a second.

A typical group for all weather defence could consist of a surveillance and acquisition radar, remote control director and two twin type A32 (M) guns. If mains supply is not obtainable then a generator is required for each gun.

The data in the table above relates to the weapons fitted to the British trailer type FV 4205, this has a double air line pressure brake system plus a mechanical hand-brake, weight of trailer without mount is about 2500kg. The wheels are removed when firing for long periods, although they can be left on should the need arise. Four screw jacks are provided (extension of 340mm) to give the platform stability when firing without the wheels.

The A32 (M) guns could also be fitted to other trailers, for example the Bofors L70. They could also be fitted on a tracked vehicle.

Employment

At the time of writing the A32 (M) is not yet in service.

A32(M) Guns on modified FV 2505D trailer

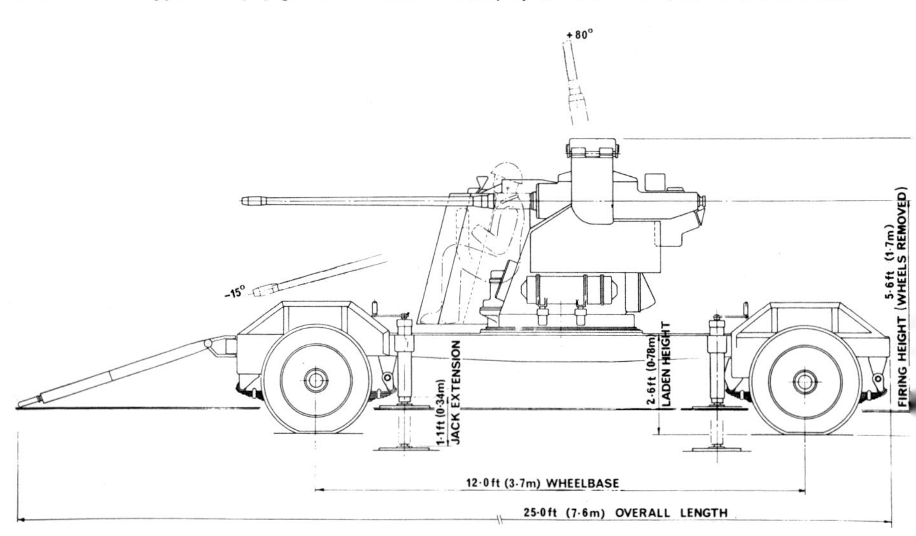

Israel

81mm Medium Mortar (SOLTAM)

This has a total weight of 58kg, maximum range is 3200m, although the effective range is 2300m. Crew is 3 men and rate of fire is 12–15rpm. This mortar can be mounted on the M-3 Half Track, in this role it is on a turntable with a 360° traverse and is used as mobile artillery in support of mechanised columns.

Two other versions are:

a) short barrelled (1150mm), range 140m–3980m, barrel weighs 11·5kg.
b) long barrelled, range 150m–4660m, barrel weighs 14·5kg.

Both versions have a baseplate weighing 13·2kg and a bipod weighing 13·2kg. Elevation is +43° to +80°.

The barrel has a smooth bore and is made of special chrome-nickel steel. The end of the barrel is threaded to accept the screwed cap of the breech plates. The ball-shaped base of the breech pieces sit in a socket in the baseplate, giving a full 360° traverse. The bipod legs support the elevating and traversing gear, and the cross-levelling gear which has both coarse and fine adjustments.

Ammunition used is the M-570, HE, length 382mm and the ML-61, HE, length 412mm. The ML-61 can only be fired at maximum charge in the long-barrelled model.

120mm Heavy Mortar (SOLTAM)

This has a weight of 360kg and a maximum range of 6500m, fires a HE bomb with a m/v of 342 m/s, the bomb has a total weight of some 13·8kg of which 1·54kg is HE, it has an auxiliary solid fuel booster. Crew is 4 men. This version is carried by an M-3 Half-Track and is used as mobile artillery in infantry brigades, i.e. one regiment of 12 to each brigade.

There is also a lightweight model weighing 108·5kg, this can be broken down into three sections: barrel weighing 38·6kg, bipod weighing 27·5kg and baseplate weighing 37kg. Each of these can be carried by one man. Range is 300–6200m.

160mm Heavy Mortar (SOLTAM)

This is the Finnish Tampella Mortar produced in Israel with many modifications by IDF designers. It has a weight of 1700kg (when on two-wheeled carriage), a range of 9300m, crew of 7 and is mounted on a modified M-4 Sherman chassis. The modifications include:

a) Gear and wheel arrangements and all such auxiliary items discarded.
b) Recoil apparatus was discarded and assembly arrangements altered.
c) A loading device has been added to give a high rate of fire.
d) Stored inside both sides of the vehicle are movable receptacles for the shells. Every shell, with fuse, is in a container that moves on a ramp—so that as each shell is fired the next is ready to be loaded. Empty cases are ejected from the rear.

Multiple Rocket Systems

Israel captured large numbers of Soviet 240mm BM-24 and M-51 (32 round) 130mm Czechoslovak rocket launchers during the 1966 campaign. These are now used by the Israeli Forces.

Israel captured large amounts of Soviet and some British (i.e. 25 Pounders) artillery pieces during the campaign, these are probably held in reserve as Israel does not use a great deal of towed artillery.

155mm Israeli Self-Propelled Gun

This is a Sherman chassis on which has been fitted the 155mm M-1950 French Howitzer, this has a range of some 17,000m. The Israelis still use the 155mm M 1950 in the towed role.

The latest Israeli self-propelled weapon is the L-33. This is a Sherman chassis with a Soltam 155mm gun/howitzer. This has a range of 21,500m. There is also a towed version of this in service with the Israeli Army, this being called the M-68 155mm gun/howitzer.

105mm Model 56 Pack Howitzer — Italy

Calibre: 105mm
Weight: 1290kg
Length: 5300mm
Length Travelling: 3650mm
Width: 1498mm
Height Travelling: 1955mm
Height Firing: 1930mm
Height Firing (Anti-Tank): 1550mm
Barrel Length: 1716mm (including muzzle brake)
Elevation: $-5° +65°$
Traverse: 36° (total)
56° (total in anti-tank role)
Rate of Fire: 8rpm
Range: 10,200m
Ammunition: HE, Smoke, AP, Target Indicating, Illuminating. 7 charges
Armour Penetration: 116mm with HEAT round
Crew: 6 (on gun)
Towing Vehicle: Britain, Volvo Bv.202 (gun on skies)
Britain, LWB Land Rover
Airportable: AB-205 helicopter—one load
Wessex helicopter—one load
C-130 aircraft can carry 2 guns and 2 towing vehicles plus crew

This weapon was designed in Italy in the early 1950s, the first prototype being built in 1956. The gun entered service with the Italian Army in 1957. Since then it has been exported all over the world by its manufacturers, Oto Melara Spa of La Spezia.

The 105mm Pack Howitzer is used extensively by airborne and mountain units. This is because the weapon was designed to be broken down into 11 loads, the heaviest of which is of 122kg. It can thus be easily transported by mule over rough terrain. The weapon can also use the American M1 ammunition as used in the M-101.

The weapon can also be used in the anti-tank role. For this role the wheels are moved forward into sockets on suspension arms. This lowers the profile of the weapon and the trails are splayed out wider. When used in the anti-tank role the weapon has a greater traverse.

When travelling the two rear sections of the trails fold up and the two main arms lock together. It is recognisable by its shield, multi-baffle muzzle brake and its folded trails whilst travelling. The German weapons have a much bigger muzzle brake. British weapons are often seen with their shields removed.

Employment

Argentina, Australia, Bangla Desh, Belgium, Canada, Ceylon, Chile, Eire, France, Great Britain (to be replaced by British 105mm Light Gun, the Italian 105mm Gun is known as the L5 (Carriage L10A1) in British use), Ghana, Germany, India, Italy, Malaysia, New Zealand, Nigeria, Pakistan, Rhodesia, Saudi Arabia, Spain, Sudan, Zambia.

German 105mm Pack Howitzer. Note its different muzzle device.

ABOVE
105mm Pack Howitzer in field role

BELOW
105mm Pack Howitzer on skis being towed by a Swedish BV.202 vehicle as used by the British Army

FOOT
105mm Pack Howitzer and Land Rover being lifted by a Royal Navy Sea King helicopter

Japan

75mm Mountain (Pack) Gun Model 94 (1934)

This was developed before World War II and is reported to be used by North Vietnam. It can be broken down into 11 units for transport by animals. Basic data is as follows:

Calibre: 75mm
Weight Travelling: 544kg
Length Travelling: 3990mm
Width Travelling: 1320mm
Elevation: −9° +45°
Traverse: 40° (total)
Maximum Range: 7000m
Rate of Fire: 10–12rpm
Ammunition: HE, AP, Shrapnel, Incendiary, Illuminating, Smoke

It has a horizontal sliding breechblock, split trails, hydropneumatic recoil mechanism, spoked wheels and the barrel has no muzzle brake.

75mm Mountain Gun Model 41 (1908)

This is recognisable by its spoked wheels, small shield, short barrel, the trails consist of two tubes connected by a cross bar, connected to this cross bar is a single tubular trail. It has a screw type breechblock and a hydrospring recoil mechanism. Basic data is as follows:

Calibre: 75mm
Weight Travelling: 544kg
Length Travelling: 3500mm
Width Travelling: 1220mm
Elevation: −18° +40°
Traverse: 6° (total)
Maximum Range: 7130m
Rate of Fire: 10rpm
Ammunition: AP, HE, hollow charge, Illuminating, Shrapnel, Incendiary, Smoke

Employment
Reported to be used by China and North Vietnam.

75mm Field Gun Model 38 (1905)

Has spoked wheels, small shield, box type trail and horizontal sliding breechblock. Basic data is as follows:

Calibre: 75mm
Weight Travelling: 1134kg
Length Travelling: 5230mm
Width Travelling: 1500mm
Height Travelling: 1600mm
Elevation: −8° +43°
Traverse: 7° (total)
Maximum Range: 8500m
Rate of Fire: 10–12rpm
Ammunition: HE, AP, Shrapnel, Incendiary, Illuminating, Smoke

70mm Howitzer Model 92 (1932)

This pre-World War II gun is easily recognisable by its small shield, solid wheels with eight holes in each and very short barrel. Has split trails and an unusual breechblock. This weapon has also been built post-war in Communist China. Basic data is as follows:

Calibre: 70mm
Weight Travelling: 220kg
Length Travelling: 2210mm
Width Travelling: 915mm
Height Travelling: 1090mm
Elevation: −10° +50°
Traverse: 45° (total)
Maximum Range: 2800m
Rate of Fire: 10rpm
Ammunition: HE, AP and Smoke

TOP RIGHT
75mm Mountain (Pack) Gun Model 94 (1934)

UPPER RIGHT
75mm Mountain Gun Model 41 (1908)

LOWER RIGHT
75mm Field Gun Model 38 (1905)

RIGHT
70mm Howitzer Model 92 (1932)

Post-War Japanese Weapons

As far as is known the only weapons manufactured are the following: American 106mm M-40 Recoilless Rifle built as the Type 60 Recoilless Rifle. 35mm Oerlikon K-63, 2×35mm Anti-Aircraft Guns have also been built in Japan.

203mm M-1955 Field Howitzers being towed by AT-T tractors

203mm M-1955 Field Howitzer Soviet Union

Calibre: 203·2mm
Weight Firing: 20,400kg
Length Travelling: 10,485mm
Width Travelling: 2996mm
Height Travelling: 2621mm
Track: 2195mm
Ground Clearance: 400mm
Barrel Length: 8800mm (including muzzle brake)
Elevation: -2° $+50^{\circ}$
Traverse: 44° (total)
Maximum Range: 29,250m
Rate of Fire: 1rpm
Ammunition: HE wt 102kg, m/v 790 m/s
Towing Vehicle: AT-T Tractor

This is one of the latest Soviet artillery weapons and is also known as the 203mm Gun-Howitzer M-1955. It has a screw type breechblock and fires bag-type, variable-charge separate loading ammunition. The normal round is HE but the weapon probably has nuclear capability.

It is recognisable by its long barrel surrounded by a cylindrical casing, the barrel has a pepperpot type muzzle brake, no shield, whilst travelling the barrel is pulled back over the split trails, i.e. out of battery. The recoil mechanism is below the barrel, the trails are of a box section. A small limber is provided, the carriage has twin disc wheels with large tyres.

Employment
Warsaw Pact Forces including Bulgaria and the Soviet Union. China and Egypt.

152mm M-1955 (D-20) Field Howitzer

Soviet Union

Calibre: 152·4mm
Weight Firing: 5650kg
Length Travelling: 8138mm
Width Travelling: 2027mm
Height Travelling: 2758mm
Ground Clearance: 250mm
Barrel Length: 5060mm
Elevation: -5° $+63^\circ$
Traverse: 90° (total) (see below)
Maximum Range: 17,300m (HE round)
Rate of Fire: 4rpm
Ammunition (Separate Loading): HE wt 43·6kg, m/v 655 m/s
APHE wt 48·8kg, m/v 600 m/s
Armour Penetration: 124mm at 1000m (APHE round)
Crew: 8–10
Towing Vehicle: AT-L or AT-S Tractor URAL-365, 6×6, Truck

The D-20 uses the same carriage as the 122mm D-74 Field Gun and has replaced the M-1943 152mm Howitzer in many units.

The weapon has a semi-automatic vertical sliding wedge type breechblock and fires case-type, variable-charge, separate-loading ammunition. In addition to the HE and APHE rounds it also fires a chemical round.

The main differences between the D-20 and the D-74 is that the D-20 has a shorter, thicker stepped barrel with a muzzle brake. Other features are the same as the D-74: recoil system above the barrel, irregular shaped sloping shield with a vertical sliding centre section, the shield overlaps the wheels. The weapon can be quickly traversed through 360° by means of a firing jack that folds up under the barrel, it has split box section trails each with a caster wheel, this folds up on top of the trail when the weapon is being towed. Two folding trail spades are also provided.

Employment
Czechoslovakia, East Germany, Hungary, India, North Vietnam, Poland, Romania, Soviet Union.

152mm M-1955 (D-20) Field Howitzer in travelling order

152mm M-1943 (D-1) Field Howitzers in travelling order

152mm M-1943 (D-1) Field Howitzer

Soviet Union

Calibre: 152·4mm
Weight Firing: 3600kg
Length Travelling: 7558mm
Width Travelling: 1994mm
Height Travelling: 1854mm
Track: 1800mm
Ground Clearance: 370mm
Barrel Length: 4207mm
Elevation: -3° $+63\frac{1}{2}^\circ$
Traverse: 35° (total)
Maximum Range: 12,400m (HE round)
Rate of Fire: 3–4rpm
Ammunition: HE wt 39·9kg, m/v 508 m/s
Semi-AP wt 51·1kg, m/v 432 m/s
Armour Penetration: 82mm at 1000m with Semi-AP round
Crew: 7
Towing Vehicle: AT-S Tractor
URAL-375, 6×6, Truck
ZIL-151, 6×6, Truck
ZIL-157, 6×6, Truck

This weapon has a screw type breech-block and fires case-type, variable-charge, separate-loading ammunition. It has the same carriage and shield as the M-1938 (M-30) 122mm Howitzer.

It is recognisable by its short barrel with a double baffle muzzle brake, recoil system above and below the barrel (hydraulic buffer above and hydropneumatic recuperator below barrel), the centre section of the shield is raised when in the firing position, top half of the shield slopes to the rear. The carriage has two large single-tyred wheels, the trails are of a box section with folding spades.

Employment
Warsaw Pact Forces including East Germany, Poland and Hungary, Albania, China, Egypt, Syria, North Vietnam.

152mm M-1938 (M-10) Howitzer Soviet Union

Calibre: 152·4mm
Weight Firing: 4150kg
Weight Travelling: 4550kg
Length Travelling: 6399mm
Width Travelling: 2096mm
Height Travelling: 1900mm
Track: 1655mm
Ground Clearance: 305mm
Barrel Length: 3700mm
Elevation: -1° $+65^\circ$
Traverse: 50° (total)
Maximum Range: 12,400m (HE round)
Rate of Fire: 4rpm
Ammunition: HE wt 39·9kg, m/v 508 m/s
Semi-AP wt 51·1kg, m/v 432 m/s
Armour Penetration: 82mm at 1000m with semi-AP round
Crew: 7
Towing Vehicle: ZIL-111

This weapon was developed before World War II, the M-1940 (M-60) 107mm Gun which appeared at the same time used the same carriage as the 152mm M-1938 (M-10). However, during the war the gun was mounted on the carriage of the M-1938 (M-30) suitably modified, it was then known as the M-1943(D-1), see separate entry for this weapon.

The M-1938 (M-10) fires case-type, variable charge, separate loading ammunition and has a screw type breechblock.

It is recognisable by its twin-tyred wheels, short barrel with a hydraulic buffer and hydropneumatic recuperator mounted below it, split trails of a box section with spades on the top of them. The shield has cut-off corners. The barrel is pulled back when travelling and the weapon is provided with a limber.

Employment
In reserve in Warsaw Pact Countries.

152mm M-1938 (M-10) Howitzer being towed by a tractor

152mm M-1937 (ML-20) Gun-Howitzer

Soviet Union

Calibre: 152·4mm
Weight Firing: 7128kg
Weight Travelling: 7930kg
Length Travelling: 7210mm (excluding limber)
Width Travelling: 2312mm
Height Travelling: 2260mm
Track: 1900mm
Ground Clearance: 315mm
Barrel Length: 4925mm (including muzzle brake)
Elevation: -2° $+65^\circ$
Traverse: 58° (total)
Maximum Range: 17,300m (HE round)
Rate of Fire: 4rpm
Ammunition: HE wt 43·6kg, m/v 655 m/s APHE wt 48·8kg
Armour Penetration: 124mm at 1000m (APHE round)
Crew: 10
Towing Vehicle: AT-S Tractor

This weapon was used during World War II and has now been replaced by the D-20 in the Soviet Army although it is still in use with reserve units as well as members of the Warsaw Pact Forces. It uses the same carriage as the 122mm M-1931/37 Field Gun. The M-1937 has a screw type breechblock and fires case-type, variable charge, separate loading ammunition, a modified version of this gun, the M-1937/43 (ML-20S) is used in the ISU-152 Assault Gun.

It is recognisable by its twin road wheels, long barrel with a multi-baffle muzzle brake, the recoil system is under the barrel, whilst travelling the barrel is pulled back over the trails, small shield behind the vertical balancing gears, split section trails of a box-section with detachable picket spades. A limber is attached to the rear of the weapon when it is being towed.

Employment
Albania, Algeria, Bulgaria, China, Cuba, Czechoslovakia, East Germany, Egypt, Finland, Hungary, North Korea, North Vietnam, Mongolia, Poland, Syria, Yugoslavia.

152mm M-1937 (ML-20) Gun-Howitzer with its limber attached

130mm Soviet Coastal Guns in Egypt

130mm Coastal Gun — Soviet Union

Calibre: 130mm
Weight Travelling: 19,000kg
Weight Firing: 16,000kg
Length Travelling: 13,000mm
Width Travelling: 2800mm
Height Travelling: 2600mm
Ground Clearance: 300mm
Barrel Length: 7600mm (including muzzle brake)
Elevation: −5° +45°
Traverse: 360°
Maximum Range: 27,000m
Rate of Fire: 5–10rpm
Ammunition: HE wt 33·4kg, m/v 930 m/s
APHE wt 33·6kg, m/v 930 m/s
Towing Vehicle: AT-T Artillery Tractor

This 130mm Coastal Gun is reported to be based on the German World War II 150mm Coastal Gun C/28, but using the M-1936 or M-1953 130mm Naval Gun.

The weapon is mounted on a mobile carriage with four sets of double tyred wheels. It is recognisable by its carriage, muzzle brake, curved shield and the travelling lock for the barrel at the rear of the carriage.

Employment
Egypt, Soviet Union.

M-1946 130mm Field Gun in travelling order with limber. Note the spades on the trails

M-1946 130mm Field Gun — Soviet Union

Calibre: 130mm
Weight Firing: 7700kg
Weight Travelling: 8450kg
Length Travelling: 11,730mm
Width Travelling: 2450mm
Height Travelling: 2550mm
Track: 2060mm
Ground Clearance: 400mm
Barrel Length: 7600mm (including muzzle brake)
Elevation: $-2\frac{1}{2}°$ $+45°$
Traverse: 50° (total)
Maximum Range: 31,000m (HE round) 1170m (APHE round)
Rate of Fire: 6–7rpm
Ammunition: HE wt 33·4kg, m/v 930 m/s APHE wt 33·6kg, m/v 930 m/s
Armour Penetration: 250mm at 1000m (APHE round)
Crew: 9
Towing Vehicle: ATS Tractor

The M-1946 was developed from the M-1936 Naval Gun and has replaced the M-1931/37 (A-19) gun in the Soviet Army. The weapon has a horizontal sliding wedge type breechblock and fires case-type, variable charge, separate loading ammunition.

The weapon is recognisable by its long barrel which has a pepperpot muzzle brake, a small shield (has also been seen without a shield), recoil system above and below the barrel, an inverted U type collar in front of the shield over the barrel, the trails are of the box type. Each trail has a detachable spade towards the rear of the trail. A two-wheeled limber is provided, when travelling the gun is pulled back until the breech is between the trails.

When used in Egypt this gun was linked to ground radar, fire control radar, fire control computer and automated command post and used in the counter battery role. It can also be linked with the Soviet 'SNAR-2' X-Band Fire-Control Radar mounted on the ATL Tractor.

A later model is the M-1954, this is slightly heavier than the M-1946, performance is similar to the M-1946.

Employment
Bulgaria, China, East Germany, Egypt, Finland, India, Iraq, Mongolia, Nigeria, North Korea, North Vietnam, Pakistan, Poland, Soviet Union, Syria, Yugoslavia.
NOTE: It is also reported that China has built this weapon and some of these have been supplied to Pakistan.

122mm M-1955 (D-74) Field Howitzer

Soviet Union

Calibre: 121·92mm
Weight Firing: 6600kg
Length Travelling: 9763mm
Width Travelling: 2027mm
Height Travelling: 2758mm
Ground Clearance: 390mm
Barrel Length: 6599mm (including muzzle brake)
Elevation: -2° $+50^\circ$
Traverse: 60° (total)
Maximum Range: 21,900m (HE round)
1200m (APHE round)
Rate of Fire: 6–7rpm
Ammunition: HE wt 25·5kg, m/v 900 m/s
APHE wt 25kg, m/v 950 m/s
Smoke, Illuminating
Armour Penetration: 230mm at 1000m (APHE round)

Crew: 10
Towing Vehicle: AT-S Tractor
AT-L Tractor
URAL-375, 6×6, Truck

The D-74 replaced the 122mm M-1931/37 (A-19) in the Soviet Army. It has a carriage similar to that of the 152mm D-20. The D-74 has a semi-automatic vertical sliding wedge breechblock and uses case-type, variable charge, separate loading ammunition. It has a circular firing jack which, with the caster wheels (one on each trail, and stowed over the trail whilst travelling), enables the weapon to be quickly traversed through 360°.

It is recognisable by its long stepped barrel and its double-baffle muzzle brake, its recoil system is above the barrel and a part of this is visible just forward of the shield, the shield has a vertically sliding centre section and the sides of the shield slope away in an irregular pattern. The trails are of a box section and split for firing, the circular firing jack is under the barrel.

Employment
Bulgaria, China, Cuba, East Germany, Egypt, Hungary, Nigeria, North Vietnam, Poland, Romania, USSR.

122mm M-1955 (D-74) Field Howitzer

122mm M-1931/37 (A-19) Field Gun

Soviet Union

Calibre: 121·92mm
Weight Firing: 7117kg
Weight Travelling: 7907kg
Length Travelling: 7870mm (excluding limber)
9450mm (overall)
Width Travelling: 2460mm
Height Travelling: 2270mm
Track: 1900mm
Ground Clearance: 335mm
Barrel Length: 5645mm
Elevation: -2° $+65^\circ$
Traverse: 58° (total)
Maximum Range: 20,800m (HE round)
900m (APHE round)
Rate of Fire: 5–6rpm
Ammunition: HE wt 25·5kg, m/v 800 m/s
APHE wt 25kg, m/v 800 m/s
Armour Penetration: 190mm at 1000m (APHE round)
Crew: 8
Towing Vehicle: AT-S Tractor
AT-T Tractor
KrAZ-214, 6×6, Truck

This weapon uses the same carriage as the 152mm M-1937 Gun/Howitzer. Wartime models had solid rubber tyres on spoked wheels, now, however, most have twin tyred wheels. It has a hydraulic recoil buffer and hydropneumatic recuperator, a screw type breechblock is fitted. The weapon has been replaced in the Soviet forces by the 122mm (D-74) and 130mm (M-46) Guns.

It is recognisable by its two vertical balancing gears, small shield behind the balancing gears, long barrel with no muzzle brake, it does, however, have a reinforcing band on the muzzle, recoil system is located under the barrel, the barrel is pulled back when in the travelling position. The box section trails are of riveted construction and are provided with detachable picket spades. A two-wheeled limber is provided for travelling purposes.

Employment
Albania, Algeria, Bulgaria, Cambodia, China, Cuba, Czechoslovakia, East Germany, Egypt, Guinea, Hungary, North Korea, North Vietnam, North Yemen, Poland, Romania, Syria, Tanzania, Yugoslavia, Somalia.

122mm M-1931/37 (A-19) Field Gun with trails split

122mm M-1938 (M-30) Field Howitzer

Soviet Union

Calibre: 121·92mm
Weight Firing: 2500kg
Length Travelling: 5900mm
Width Travelling: 1975mm
Height Travelling: 1820mm
Track: 1600mm
Ground Clearance: 330mm
Barrel Length: 2800mm
Elevation: -3° $+63\frac{1}{2}^\circ$
Traverse: 49° (total)
Maximum Range: 11,800m (HE round) 630m (HEAT round)
Rate of Fire: 5–6rpm
Ammunition: HE wt 21·8kg, m/v 515 m/s HEAT wt 14·1kg, also Smoke, Chemical and Illuminating rounds
Armour Penetration: 200mm (HEAT round)
Crew: 8
Towing Vehicle: ZIL-151, 6×6, Truck BUCEGI, 4×4, Truck (Romania) Valmet Terra, 4×4 (Finland)

This was used by the Soviet Army during World War II and was built in large numbers, it is still an effective weapon although it is being replaced by the 122mm D-30 weapon.

The 122mm M-1938 has a screw type breechblock and uses case-type, variable charge, separate loading ammunition. The hydraulic buffer is located below the barrel and the hydropneumatic recuperator is above the barrel, no muzzle brake is fitted. The weapon has a shield, the centre of which is raised when in the firing position. The carriage has box section trails with folding spades. It can be fired without spreading the trails, although in this case it has a limited traverse of only 1·5°.

The Bulgarian Army uses M-1938s with metal spoked type wheels, the East German Army has some M-1938s with much smaller wheels and wider tyres.

Employment
Albania, Algeria, Bulgaria, China, Cuba, Czechoslovakia, East Germany, Egypt, Finland, Hungary, Iraq, North Korea, North Vietnam, Poland, Romania, Soviet Union, Syria, Yugoslavia.

Russian 122mm M-1938 (M-30) Howitzers of the East German Army.

ABOVE
Soviet 122mm M-1938 (M-30) Howitzers being deployed, the trucks are ZIL-157s

LEFT
East German 122mm M-1938 (M-30) Howitzers with the new type wheels and tyres

122mm D-30 Howitzer — Soviet Union

Calibre: 121·92mm
Weight Firing: 3150kg
Length Travelling: 5400mm
Width Travelling: 1950mm
Height Travelling: 1660mm
Track: 1850mm
Barrel Length: 4785mm (including muzzle brake)
Elevation: −7° +70°
Traverse: 360°
Maximum Range: 16,000m (HE round) 1000m (HEAT round)
Rate of Fire: 7–8rpm
Ammunition: HE wt 21·8kg, m/v 690 m/s HEAT wt 14·1kg, m/v 740 m/s
Armour Penetration: 460mm (HEAT round)
Crew: 7
Towing Vehicle: AT-P Artillery Tractor
ZIL-157, 6×6, Truck
GAZ-63, 4×4, Truck
URAL-375D, 6×6, Truck

The D-30, full designation being 122mm M-1963 (D-30) Quick Firing Field Gun/Howitzer. It is one of the newer additions to the Soviet range of artillery and is replacing the M-1938 (M-30) Gun. It was first seen in 1967.

Early versions of this weapon reportedly include the M-1946 (D-13) and M-1949 (D-24).

The weapon has a semi-automatic vertical sliding wedge breechblock, and fires case-type, variable charge, separate loading ammunition, the case is longer as it has a larger propelling charge and therefore has a higher muzzle velocity.

The D-30 is towed muzzle first, some have the towing eye lug underneath the muzzle brake and some have it behind the muzzle brake, whilst travelling the three trail legs are clamped under the barrel. When in firing position the crew lower the central jack, raising the wheels off the ground, and then move the outer trails through 120°. The weapon then settles on its trails and can therefore be quickly traversed through 360°. This makes the gun very useful in the anti-tank role.

The weapon is recognisable by its small shield, muzzle brake, large recoil system above the barrel, and its trails.

Employment
Cuba, Czechoslovakia, East Germany, Egypt, Finland, Hungary, Poland, Soviet Union, Syria.

D-30s being used in the anti-tank role

ABOVE
122mm D-30 Howitzer in firing position. Note that the wheels are off the ground.

BELOW
122mm D-30 Howitzer in travelling position. This photograph clearly shows the trail legs under the barrel.

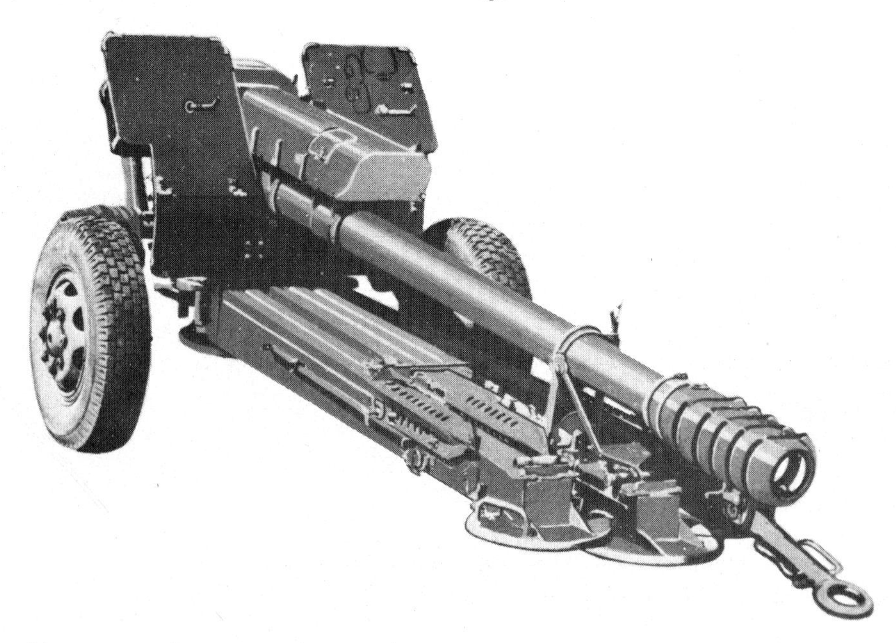

107mm B-11 Recoilless Gun Soviet Union

Calibre: 107mm
Weight Firing: 240kg
Length Travelling: 3560mm
Width Travelling: 1450mm
Height Travelling: 900mm
Track: 1250mm
Ground Clearance: 320mm
Barrel Length: 3383mm
Elevation: -10° $+45^\circ$
Traverse: 35°
Maximum Horizontal Range: 6650m (HE round)
Maximum Effective Range: 450m HEAT round)
Rate of Fire: 5–6rpm
Ammunition: HE wt 8·5kg, m/v 375 m/s HEAT wt 7·5kg, m/v 400 m/s
Armour Penetration: 380m (HEAT round) at 90° angle
Crew: 5
Towing Vehicle: ZIL-157, 6×6, Truck UAZ, 4×4, Truck

The B-11 is no longer in front line service with the Soviet Army but is still used in many other countries. It is issued on the scale of six per battalion.

The B-11 is a smoothbore recoilless weapon and fires fin stabilised HE and HEAT rounds and has an open type swinging breechblock. In some models the breech has been enlarged and covered with a grill or jacket to protect the crew from the hot chamber.

The wheels can be removed and the gun fired with or without its wheels. It is towed muzzle first or can easily be manhandled.

The B-11 is also known as the RG-107 (RG=Rückstrassfreies Geschütz). It is easily recognisable by its carriage which has the forward leg of the tripod fastened just beneath the tube when in the travelling position and the fixed tubular trails at the rear.

Employment

Bulgaria, Cambodia, China, East Germany, Egypt, North Korea, North Vietnam.

107mm B-11 being loaded. Note the breech.

M-1944 (D-10) Field Anti-Tank Gun

Soviet Union

Calibre: 100mm
Weight Firing: 3460kg
Length Travelling: 9370mm
Width Travelling: 2150mm
Height Travelling: 1500mm
Track: 1800mm
Ground Clearance: 300mm
Barrel Length: 6069mm (including muzzle brake)
Elevation: -5° $+45^\circ$
Traverse: 58° (total)
Maximum Range: 21,000m (HE round)
Rate of Fire: 8–10rpm
Ammunition (Fixed): HE wt 15·7kg, m/v 900 m/s
APHE wt 15·9kg, m/v 1000 m/s
Armour Penetration: 185mm at 1000m (APHE round)
380mm at any range (HEAT round)
Crew: 6–8
Towing Vehicle: AT-P Tractor
URAL-375D, 6×6, Truck

This weapon entered service with the Soviet Army towards the end of World War II and was based on the 100mm M-1939 Anti-Aircraft Gun. Since the war it has been replaced by the M-1955 100mm weapon. A modified version of this gun is used in the T-54 tank (D-10T) and SU-100 Assault Gun (D-10T). The M-1944 has a semi-automatic vertical sliding wedge type breechblock.

It is recognisable by its double baffle muzzle brake, long barrel, recoil system below the barrel, dual wheels (the wheels have five triangular holes in them), the shield slopes to the rear, the sides of the shield are angled to the side and overhang the wheels. On the front of each side of the shield is a stowage bin.

Employment
Bulgaria, China, East Germany, Egypt, Hungary, India, Mongolia, North Korea, North Vietnam, Poland, Romania, Somalia

M-1955 (T-12) 100mm Field Anti-Tank Gun

Soviet Union

Calibre: 100mm
Weight Firing: 3000kg
Length Travelling: 8717mm
Width Travelling: 1585mm
Height Travelling: 1890mm
Track: 1200mm
Ground Clearance: 350mm
Barrel Length: 6126mm (including muzzle brake)
Elevation: -5° $+45^\circ$
Traverse: 60° (total)
Maximum Range: 21,000m (HE round)
Rate of Fire: 7–8rpm
Ammunition: HE wt 15·7kg, m/v 900 m/s
APHE wt 15·9kg, m/v 1000 m/s
Armour Penetration: 185mm at 1000m (APHE round)
380mm with HEAT round
Crew: 7
Towing Vehicle: AT-P Tractor
ZIL-157, 6×6, Truck
URAL-375D, 6×6, Truck

The M-1955 replaced the M-1944 100mm anti-tank gun. The M-1955 has a vertical sliding, semi-automatic wedge breech-block and can fire APHE, HE or HEAT rounds.

It is recognisable by its single road wheels, long stepped barrel, this has a pepperpot muzzle brake. The recoil cylinders are on top of the barrel behind the shield. The shield has a flat front with sloping side wings. The split trails are of a box section and a single caster wheel is carried folded on top of the trails whilst travelling. The M-1955 can also be fitted with an infra-red sight for night firing.

There is also a modified version of the M-1955 gun in service. This has a larger breech ring and has a pepperpot muzzle brake that is only slightly larger in diameter than the barrel, this has the designation T-12 and it is reported that it has an elevation of -10° to $+20^\circ$.

Employment
China, Bulgaria, Czechoslovakia, East Germany, Egypt, Hungary, India, Mongolia, North Korea, North Vietnam, Poland, Romania, Somalia, Soviet Union.

TOP
Late model M-1955 100mm ATG being towed by ZIL-131 trucks. Note the caster wheel over the trail of the gun in the background.

ABOVE
100mm M-1944 (D-10) Field Anti-Tank Gun. Note the dual tyres.

Soviet SD-44 on the move. Overhead are MIL Mi-6 helicopters.

SD-44 85mm Auxiliary Propelled Gun

Soviet Union

Calibre: 85mm
Weight Firing: 2250kg
Length Travelling: 8220mm
Width Travelling: 1780mm
Height Travelling: 1420mm
Track: 1434mm
Ground Clearance: 350mm
Barrel Length: 4693mm (including muzzle brake)
Elevation: -7° $+35^\circ$
Traverse: 54° (total)
Maximum Range: 15,650m
Rate of Fire: 15–20rpm
Ammunition: HE wt 9·5kg, m/v 792 m/s
APHE wt 9·3kg, m/v 792 m/s
HVAP wt 5·0kg, m/v 1030 m/s
Armour Penetration: 102mm at 1000m (APHE round)
130mm at 1000m (HVAP round)
Crew: 5–7
Towing Vehicles (when required):
ZIL-157, 6×6, Truck
BTR-152 APC
AT-P Artillery Tractor

This is the D-44 Field Gun fitted with a M-72, two-cylinder petrol engine of 14hp, this gives the gun a road speed of 25kmph. or a cross-country speed of 8–10kmph. The weapon can also be towed. Full designation is 85mm APAT SD-44 Auxiliary Self-Propelled Anti-Tank Gun.

The weapon is recognisable by its engine, steering column and driver's seat mounted on the left trail. On the right trail is an ammunition container. The trails are hollow and carry the fuel for the engine. The steering wheel and the trail wheel are folded when the gun is in firing position, this enables the left trail spade to dig into the ground. When being towed the third wheel is slung between the trail legs. A folding ramrod is mounted either side of the shield.

Employment

Albania, Bulgaria, Cuba, Czechoslovakia, East Germany, Hungary, Poland, Romania, Soviet Union.

85mm M-1945 (D-44) Field Gun Soviet Union

Calibre: 85mm
Weight Firing: 1725kg
Length Travelling: 8340mm
Width Travelling: 1780mm
Height Travelling: 1420mm
Track: 1434mm
Ground Clearance: 350mm
Barrel Length: 4693mm
Elevation: -7° $+35^\circ$
Traverse: 54° (total)
Maximum Range: 15,650m (HE round)
Rate of Fire: 15rpm
Ammunition: HE wt 9·5kg, m/v 792 m/s
APHE wt 9·3kg, m/v 792 m/s
HVAP wt 5·0kg, m/v 1030 m/s
Armour Penetration: APHE 102mm at 1000m
HVAP 130mm at 1000m
Crew: 8
Towing Vehicle: BTR-152, 6×6, APC
ATP Tractor
ZIL-157, 6×6, Truck

This weapon is also known as the 85mm M-1945 Anti-Tank Gun, 85mm Divisional Gun D-44 or 85mm M-1945 (D-44 and D-48) Medium Anti-Tank Gun. The gun is the same as that in the T-34/85 medium tank and the M-1939 and M-1944 Anti-Aircraft Guns. The gun has a semi-automatic, vertical sliding wedge breechblock.

The D-44-N has an infra-red sight mounted on the top of the shield. Another version is the SP-44 self-propelled (auxiliary) which is described separately.

Its recognition features are its long barrel with a double baffle muzzle brake, tubular split trails, recoil system (hydraulic buffer and hydropneumatic recuperator) at the rear of the shield. The shield has a top with regular curves and its sides slope to the rear.

Employment
Albania, Algeria, Bulgaria, China, Cuba, East Germany, Egypt, Guinea, Hungary, North Korea, North Vietnam, Poland, Romania, Soviet Union.

85mm M-1945(D-44) ATG in firing position

82mm B-10 in the firing position with tripod

82mm Recoilless Anti-Tank Gun B-10

Soviet Union

Calibre: 82mm
Weight Firing: 72kg
Weight Travelling: 85kg
Length Travelling: 1910mm
Width Travelling: 714mm
Height Travelling: 673mm
Barrel Length: 1660mm (overall)
Elevation: $-20° +35°$
Traverse: 360°
Maximum Range: 4400m (HE round)
500m (HEAT round effective)
Rate of Fire: 5–6rpm
Ammunition: HE projectile wt 4·5kg, m/v 320 m/s
HEAT projectile wt 3·6kg, m/v 322 m/s
Armour Penetration: 240mm (HEAT round) armour at 90°
Crew: 4

This recoilless weapon entered service in 1950 and was developed from the earlier SPG-82. Ammunition is loaded from the rear, the firing mechanism is on the right hand side of the barrel and is in the form of a pistol grip. Optics are on the left hand side.

It has a two-wheeled carriage and is towed by its barrel. A bar-type hand grip is attached to the barrel to assist handling and the weapon can be pulled by two or three men. It can be fired with or without its wheels, in the latter case it is supported by its tripod.

The B-10 (Bezotkatnojieordie-10) can also be carried on the rear deck of the BTR-50 APC. The ammunition is fin stabilised, barrel is smooth. The weapon is known as the RG-82 in East Germany.

Employment
Bulgaria, China, East Germany, Egypt, North Korea, North Vietnam, Pakistan, Syria.

M-1942 (ZIS-3) 76mm Gun — Soviet Union

Calibre: 76·2mm
Weight Firing: 1150kg
Length Travelling: 6095mm
Width Travelling: 1645mm
Height Travelling: 1375mm
Track: 1400mm
Ground Clearance: 315mm
Barrel Length: 3455mm (including muzzle brake)
Elevation: $-5°$ $+37°$
Traverse: 54° (total)
Maximum Range: 13,290m (HE round)
Rate of Fire: 15rpm
Ammunition: HE wt 6·2kg, m/v 680 m/s
APHE wt 6·5kg, m/v 655 m/s
HVAP wt 3·1kg, m/v 965 m/s
HEAT wt 4·0kg, m/v 325 m/s
Armour Penetration: 69mm at 500m (APHE round)
92mm at 500m (HVAP round)
120mm at any range (HEAT round)
Crew: 6–7
Towing Vehicle: BTR-152 APC, 4×4 or 6×6, Truck

This weapon entered service in 1942, earlier versions included the M-1936, M-1939 and M-1941, the weapon is no longer in service with the Soviet Army, although it may be held in reserve.

The carriage is a modified 57mm carriage with tubular trails and spades on each end. It has a vertical sliding breech-block.

Its recognition features are its double baffle muzzle brake, square type shield that projects over the wheels, tubular split trials and hydraulic buffer and hydropneumatic recuperator above and below the barrel. It is built in Communist China under the designation 57mm Anti-Tank Gun Type 55.

Employment
Albania, Cambodia, China, Cuba, Bulgaria, Czechoslovakia, East Germany, Egypt, Finland, Ghana, Hungary, Indonesia, Nigeria, North Korea, North Vietnam, North Yemen, Poland, Romania, Tanzania, Yugoslavia.

M-1942 (ZIS-3) 76mm Gun being towed by a 4×4 truck

Photograph shows a 57mm Auxiliary Propelled Anti-Tank Gun being towed by an East German Robur LO 1800 A series 4×4 truck

57mm Auxiliary Propelled Anti-Tank Gun

Soviet Union

Calibre: 57mm
Weight Firing: 1250kg
Length Travelling: 6212mm
Width Travelling: 1800mm
Height Travelling: 1220mm
Track: 1400mm
Barrel Length: 4070mm
Elevation: −5° +15°
Traverse: 56° (total)
Maximum Range: 7000m (HE round)
Rate of Fire: 15–20rpm
Armour Penetration: 106mm at 500m (APHE round)
140mm at 500m (HVAP round)
Ammunition: HE wt 2·8kg, m/v 706 m/s
APHE wt 3·1kg, m/v 990 m/s
HVAP wt 1·8kg, m/v 1255 m/s
Crew: 5–6

This is also known as the 57mm M-1955 Light Anti-Tank Gun (Auxiliary Self-Propelled). On the right hand trail is mounted a M-72, two-cylinder, 14hp auxiliary petrol engine, this enables the gun to propel itself at speeds of up to 40kmph on roads. Also on the right hand trail is the steering wheel and the driver's seat. The trails are of a box section with spades on each end, the third wheel is also behind the spades. The weapon can also be towed. When being towed the third wheel is swung up above the trails.

The weapon has a semi-automatic vertical sliding wedge type breechblock, infra-red sighting equipment can be fitted if required.

This gun can be recognised by the engine on the right hand trail, drive shaft from the engine to the two wheels. Shield has rounded edges and is lower than the M-1943 (ZIS-2), long barrel with a double baffle muzzle brake and a recoil cylinder underneath it. Each trail also has an ammunition box on it.

Employment
Albania, Cuba, Czechoslovakia, East Germany, Guinea (reported), Hungary, Poland and Romania.

M-1943 (ZIS-2) 57mm Anti-Tank Gun

Soviet Union

Calibre: 57mm
Weight Firing: 1150kg
Length Travelling: 6795mm
Width Travelling: 1700mm
Height Travelling: 1370mm
Track: 1400mm
Ground Clearance: 315mm
Barrel Length: 4160mm
Elevation: -5° $+25^{\circ}$
Traverse: 56° (total)
Maximum Range: 8400m (HE round)
Rate of Fire: 20–25rpm
Ammunition: HE wt 2·8kg, m/v 706 m/s
APHE wt 3·1kg, m/v 990 m/s
HVAP wt 1·8kg, m/v 1270 m/s
Armour Penetration: 106mm at 500m with APHE round
140mm at 500m with HVAP round
Crew: 5–7
Towing Vehicle: BTR-152, 6×6, APC
GAZ-69, 4×4, Truck
BTR-50P (Carried in)

This is also known as the 57mm M-1943 (ZIS-2) Light Anti-Tank Gun or Chinese Communist 57mm Anti-Tank Gun Type 55. It was developed from the earlier M-1941 Anti-Tank Gun and uses the same carriage as the M-1942 (ZIS-3) 76mm Gun. The weapon has a semi-automatic vertical sliding wedge breechblock and fires the same ammunition as the ASU-57 Self-Propelled Anti-Tank Gun and the 57mm Auxiliary Propelled Anti-Tank Gun. An infra-red night sight can also be fitted.

It is recognisable by its recoil system above and below the barrel, tubular split trails, no muzzle brake, straight topped shield (in recent years, however, some of these weapons have been observed with a wavy top to their shields), fixed spades and a limber loop.

Employment
Albania, China, Cuba, Bulgaria, Czechoslovakia, East Germany, Egypt, Hungary, North Korea, Poland, Romania, Yugoslavia.

57mm M-1943 Anti-Tank Gun being unloaded from a helicopter

M-1942 45mm Anti-Tank Gun captured in Korea

M-1942 45mm Anti-Tank Gun — Soviet Union

Calibre: 45mm
Weight Firing: 570kg
Length Travelling: 4885mm
Width Travelling: 1634mm
Height Travelling: 1200mm
Track: 1400mm
Ground Clearance: 254mm
Barrel Length: 3087mm
Elevation: -8° $+25^\circ$
Traverse: 30° (left and right)
Maximum Range: 4400m (HE round)
Rate of Fire: 25rpm
Ammunition: HE wt 2·1kg, m/v 343 m/s
APHE wt 1·4kg, m/v 820 m/s
HVAP wt 0·9kg, m/v 1070 m/s
Armour Penetration: 66mm at 500m (HVAP round)
Crew: 6
Towing Vehicle: GAZ-67, UAZ-69, 4×4, Truck

The M-1942 gun uses the same carriage as the earlier M-1937, this had a shorter barrel and spoked wheels.

The M-1942 is recognisable by its split tubular trails, each with a spade, recoil system under the long barrel which has no muzzle brake, and the shield with an irregular top and wings.

The weapon has been seen with two types of tyred wheels, one has plain wheels (as in the photograph) and the other has triangular holes. The gun has a semi-automatic vertical sliding wedge type breechblock and a hydraulic buffer and a spring recuperator.

Employment
Albania, North Korea.

240mm Heavy Mortar M-1953 being loaded

240mm Heavy Mortar M-1953 Soviet Union

Calibre: 240mm
Weight in Firing Position: 3610kg
Length in Travelling Position: 6510mm
Width in Travelling Position: 2490mm
Height in Travelling Position: 2210mm
Barrel Length: 5340mm (22·4 calibres)
Elevation: +45° +65°
Traverse: 17° (total)
Rate of Fire: 1rpm
Ammunition: Wt 100kg, m/v 362 m/s (HE round)
Maximum Range: 10,000m
Minimum Range: 1500m
Crew: 8
Towing Vehicle: AT-P Tractor
AT-L Tractor
AT-S Tractor

This mortar is the largest mortar used by the Soviet Army and is also called the M-240. There have been reports in recent years, however, that this weapon is no longer in service with the Soviet Army.

It is normally towed muzzle first by a tracked vehicle. A two-wheeled trolley is used to assist loading the ammunition. It is breech loaded and fired by a lanyard. To load the weapon the barrel is swung into the horizontal position in a similar way to the 160mm mortar M-1953. The 240mm M-1953 is reported to have nuclear capability.

It is recognisable by its very large circular baseplate (213cm in diameter), which has six strengthening pieces, elevating handwheel on the left of the barrel, two vertical cylinders either side of the barrel, collar around the barrel in which the trunions are located and the ground support plate and pickets below and astride the barrel.

Employment
Bulgaria, China, Romania, Soviet Union.

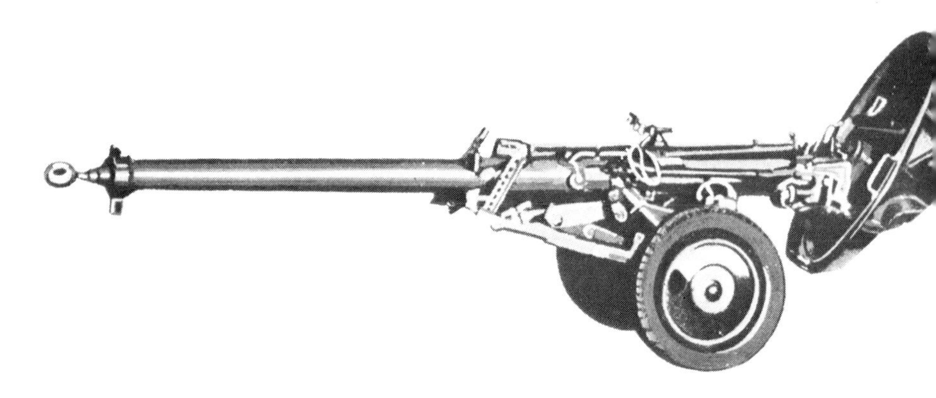

160mm M-1953 Mortar. Note towing lug on muzzle.

160mm M-1953 Heavy Mortar   Soviet Union

Calibre: 160mm
Weight in Firing Position: 1300kg
Weight in Travelling Position: 1470kg
Length in Travelling Position: 4860mm
Width in Travelling Position: 2030mm
Height in Travelling Position: 1690mm
Track: 1790mm
Ground Clearance: 360mm
Length of Barrel (Overall): 4550mm
Elevation: +50° to +80°
Traverse: 24° (total)
Rate of Fire: 2–3rpm
Maximum Range: 8040m
Minimum Range: 750m
Ammunition: HE wt 41·5kg, m/v 345 m/s
Crew: 7–9
Towing Vehicle: BTR-152, 6×6, APC
ZIL-157, 6×6, Truck

This weapon was first seen in 1953 and has a much longer barrel than the M-1943 and a very large (152·4cm) circular baseplate. To load the mortar the breech is released from the baseplate and swung into a horizontal position, after the round has been loaded it is returned to its original position. It is fired by a lanyard which is attached to the trigger. It is built in China as the Type 60.

It is recognisable by its balancing cylinders below the handwheels and the support arms and anchor pickets below and astride the barrel. It is towed by its muzzle on a two-wheeled carriage, the wheels have five holes in them, although different types of wheels have been seen.

The M-1953 is believed to be obsolete in the Soviet Army although it has been reported in mountain units, however it is rather heavy for this role. It is often called the M-160 Heavy Mortar.

Employment
United Arab Republic, China.

* NOTE: *Some reports have indicated that there is a Soviet 120mm mortar that fires a rocket-assisted projectile, it is called PRM-65.*

160mm Mortar M-1943 in travelling position

160mm Mortar M-1943

Soviet Union

Calibre: 160mm
Weight Firing: 1080kg
Length Firing: 4042mm
Width Firing: 1786mm
Height Firing: 1414mm
Ground Clearance: 308mm
Barrel Length: 3230mm
Elevation: +45° to +85°
Traverse: 25°
Rate of Fire: 3rpm
Maximum Range: 5150m (HE round)
Projectile Weight: 40·50kg (HE round)
Crew: 4–8
Towing Vehicle: Light Truck

This was introduced into the Soviet Army in 1943 and at that time was the largest mortar in service with the Soviet Army. It is also known as the 160mm Heavy Mortar M-1943. The weapon is no longer in service with the Soviet Army having been replaced by more modern mortars such as the M-1953.

The weapon is of the smooth-bore breech-loading type. The barrel being swung horizontal for loading purposes. It is recognisable by its short barrel, large (152cm) diameter baseplate with small carrying handles, elevating and traversing handwheels are on the left hand side of the barrel at 45°, box-shaped cradle below the barrel, the single-tyred wheels have triangular holes in them. The weapon is towed muzzle first.

Employment
Albania, China, Czechoslovakia, East Germany, Egypt, North Korea, North Vietnam, Syria.

120mm Mortar M-1943 being unloaded from its trailer

120mm M-1938 and M-1943 Medium Mortars
Soviet Union

Calibre: 120mm
Weight Firing: 274kg
Barrel Length: 1854mm (without muzzle attachment)
Elevation: +65° +80°
Traverse: 8° (total)
Maximum Range: 5700m
Minimum Range: 460m
Rate of Fire: 12–15rpm
Ammunition: Projectile wt 15·4kg, m/v 272 m/s
Crew: 6
Towing Vehicle: ZIL-157, 6×6, Truck

This mortar is used by most members or the Warsaw Pact and can be found both at battalion level, where it has replaced the 82mm Mortar in some cases, or at regimental level.

The mortar is normally carried horizontally on a two-wheeled tubular trailer, with the baseplate at the rear end. The towing vehicle carries the crew and ammunition. It can also be broken down into three loads for animal transport.

The projectile can be either (HE) of cast or wrought iron, cast iron has a better effect but a slightly shorter range. The weapon can also fire smoke and incendiary rounds. The barrel is smooth and a device can be fitted to prevent double loading.

The weapon is recognisable by its circular baseplate which is 106cm in diameter with webbed fins underneath, two legs that are joined by a spring-loaded chain across the bottom, the right leg has an anti-cant device attached to it, twin recoil cylinders extend from the traversing gear to near the travelling clamp.

The M-1943 is recognisable by the greater length of the recoil cylinders. Both the M-1938 and M-1943 fire the same round and can also be fired by trigger as well as normal drop firing.

NOTE: The M-1943 has also been built in China under the designation Type 53 Mortar.

Employment
Albania, China, Czechoslovakia, East Germany, Egypt, Iraq, North Korea, North Vietnam, Romania, Soviet Union, Syria, Yemen, Yugoslavia.

NOTE: The M-1943 is used by Austria under designation of M-60.

107mm M-1938 Mortar
Soviet Union

Reported to be used by China, North Korea and North Vietnam.

ABOVE
120mm M-1938 Mortars

BELOW
120mm M-1943 Mortar. The device on the end of the barrel is to prevent double loading, a similar device being used on the 82mm Mortars.

82mm Light Mortar M-1937, M-1941 and M-1943 Soviet Union

Calibre: 82mm
Weight Firing: 56kg
Barrel Length: 1220mm
Elevation: +45° +85°
Traverse: 6°
Maximum Range: 3040m
Minimum Range: 100m
Rate of Fire: 15–25rpm
Ammunition: HE, projectile wt 3·315kg
HE, projectile wt 3·087kg
Smoke, projectile wt 3·454kg
Also reported is white phosphorous booster shell with a range of 3800m
Illuminating—range 2000m
Crew: 4–5
NOTE: *Above data relates to the M-1943.*

82mm M-1937 Mortar being used by the East German Army

M-1937

The M-1936 82mm Mortar was copy of the M-1917/31 Stokes-Brandt Mortar and was developed into the M-1937. It is a smooth bore, muzzle loaded mortar and can be broken down into three parts for easy transportation, barrel, baseplate and bipod. The bipod is attached to the barrel near the muzzle end and has two recoil cylinders. Baseplate is circular, 0·58m in diameter and is of welded construction.

M-1941

The M-1941 can be broken down or towed by its two metal disc type wheels attached to the bipod, these are removed for firing. The bipod is attached midway down the barrel and a recoil cylinder is mounted between the bipod and the baseplate. Weight travelling is 50·3kg, weight firing 45kg, barrel weight 19·5kg, baseplate weight 19kg.

M-1943

May be carried or towed. The wheels are not removed for firing, there are support legs on the axle between the two wheels. The recoil cylinder is in a similar position to the M-1941 model.

Sights fitted are the MP-1 optical sight and the MP-82 collimeter sight, these also being fitted to the 120mm mortars. The mortars can also be fitted with a muzzle safety device which prevents double loading. It is also reported that these 82mm mortars will fire the 81mm rounds used in British, United States, French and Japanese mortars. The M-1937 is also built in Communist China under the designation Type 53, it is also reported that they are, or have been, built in North Korea and East Germany.

The M-1937, M-1941 and M-1943 are no longer used by the Soviet Army.

Employment
Albania, Bulgaria, China, Cambodia, Cuba, Cyprus, Czechoslovakia, Congo, Egypt, Ghana, Indonesia, East Germany, North Korea, North Vietnam, North Yemen, Iraq, Syria, Yugoslavia.

25mm M-1940 Anti-Aircraft Gun Soviet Union

This is reported to be in limited use with the North Vietnamese. There are two versions, either a single barrelled version mounted on a four-wheeled carriage with a shield or a twin version mounted on a truck. Basic data is: calibre 25mm, rate of fire 100rpm, traverse 360°, elevation −10° +85°. Maximum range in the anti-aircraft role is 4500m and in the ground role 6400m. Its effective range, however, is much less.

76mm M-1938 Anti-Aircraft Gun Soviet Union

This may be in limited use in North Vietnam and other Soviet satellites. It uses the same carriage as the 85mm M-1939 (KS-12) Anti-Aircraft Gun. Basic data is: calibre 76·2mm, rate of fire 12–15rpm, traverse 360°, elevation −3° +82°, maximum anti-aircraft range is 9500m and in the ground role 14,000m. Fires HE and APHE rounds.

76mm Mountain Gun M-1938 Soviet Union

This is reported to be used in North Vietnam. It can be broken down into 10 loads for easy transportation. Basic data is: calibre 76·2mm, weight 786kg, range 10,000m with HE rounds (m/v 503 m/s), rate of fire 14rpm, elevation −8° +70°, traverse 10°.

76mm Howitzer M-1943 Soviet Union

Yet another old weapon reported to be used by North Vietnam. This is mounted on a modified carriage of the 45mm M-1942 Anti-Tank Gun. Basic data is: calibre 76·2mm, range 4200m, weight 600kg, traverse 60°, elevation −8° +25°. Rate of fire 14rpm.

76mm M-1902/30 Gun Soviet Union

Reported to be in use in North Vietnam. This very old weapon may be seen with two types of barrel, one is 2·90m long and the other 2·14m. Basic data is: calibre 76·2mm, range 12,000m (m/v 717 m/s), weight 1350kg, rate of fire 8rpm, elevation −5° +37° and traverse 2°.

105mm M-1941 Field Gun Soviet Union

This is used by Cuba, Guinea and North Vietnam and in a modified version by Finland (m/61).

76mm M-1927 Howitzer Soviet Union

Calibre: 76·2mm
Weight Firing: 780kg
Elevation: −6° +25°
Traverse: 6°
Rate of Fire: 14rpm
Range: 8570m (projectile wt is 6·2kg and m/v 387 m/s)

This weapon is no longer in front line use with any of the Warsaw Pact Forces but it may be in reserve or used as a training or saluting weapon. It is recognisable by its very large wheels, these wheels can be either wooden (with steel or rubber tyres) or steel disc type with run-flat tyres.

130mm M-1955 (KS-30) Heavy Anti-Aircraft Gun

Soviet Union

Calibre: 130mm
Weight Firing: 24,900kg
Weight Travelling: 29,500kg
Length Travelling: 11,520mm
Width Travelling: 3033mm
Height Travelling: 3048mm
Track: 2388mm
Ground Clearance: 408mm
Barrel Length: 8412mm
Elevation: $-5°$ $+80°$
Traverse: 360°
Maximum Range: 29,000m (Horizontal)
21,900m (Vertical)
17,000m (Vertical-effective)
Rate of Fire: 10–12rpm
Ammunition: HE wt 33·4kg, m/v 945 m/s
Crew: 12
Towing Vehicle: AT-T Tractor
URAL-375, 6×6, Truck

This weapon is similar in many respects to the American 120mm anti-aircraft gun. It was first seen in 1955 but has been replaced by SAM systems in the Soviet Union and Warsaw Pact Forces.

The M-1955 fires fixed charge separate ammunition and has a semi-automatic horizontal sliding wedge type breech-block. It is also provided with an automatic fuse setter and an automatic rammer. It is used in conjunction with the PUAZO-30 director and fire control radar SON-30. The new fire control system is the 'Fire-Can'. The gun is mounted on a carriage which has four double tyred wheels, out-riggers (folded up when in the travelling position) are mounted on the front, rear and either side of the carriage. It has no shield.

Recognition features are its long stepped barrel without a muzzle brake, the barrel is held in place whilst in the travelling position by a lock at the rear of the car-riage. The firing platform (at the rear of the gun) folds up at about 45° whilst travelling. Recoil system below the barrel and the long balancing gears under the barrel below the recoil system.

Employment
North Vietnam.

130mm M-1955 (KS-30) Heavy Anti-Aircraft Gun

100mm M-1949 (KS-19) Heavy Anti-Aircraft Gun

Soviet Union

Calibre: 100mm
Weight Travelling: 19,450kg
Length Travelling: 9238mm
Width Travelling: 2286mm
Height Travelling: 2201mm
Track: 2165mm
Ground Clearance: 305mm
Barrel Length: 5742mm
Elevation: -3° $+85^\circ$
Traverse: 360°
Maximum Range: 21,000m (horizontal range)
15,400m (vertical range)
Rate of Fire: 15–20rpm
Ammunition: HE wt 15·7kg, m/v 900 m/s
APHE weight 15·9kg, m/v 1000 m/s
Armour Penetration: 185mm at 1000m (APHE round)
Crew: 9
Towing Vehicle: AT-S Tractor
URAL-375, 6×6, Truck
Praga V3S, 6×6, Truck (Czechoslovak).

This replaced the older 85mm M-1939 (KS-12) and M-1944 (KS-18), but it is no longer in front line service with any of the Warsaw Pact Forces.

It has a semi-automatic horizontal sliding wedge type breechblock, an automatic fuse setter and a power rammer. Effective range in the anti-aircraft role is 13,000–14,000m. The gun is usually used with the PUAZO-6/19 director and the SON-9 or SON-9A fire control radar. The FANSONG-E is replacing the SON-9 and SON-9A fire control radar.

The gun is mounted on a four-wheeled carriage with a rectangular firing platform, the shield folds down, wheels are removed when in the firing position, folding outriggers are provided. A travel lock is provided on the rear of the carriage.

Recognition features are its long stepped barrel with a multi-baffle muzzle brake, shield in front, twin balancing gears, recoil system above and below the barrel. On the left of the platform is a bracket to carry one round of ammunition for ready use. Above and to the left of the breech is the loading tray, the power gear is on the right side of the gun.

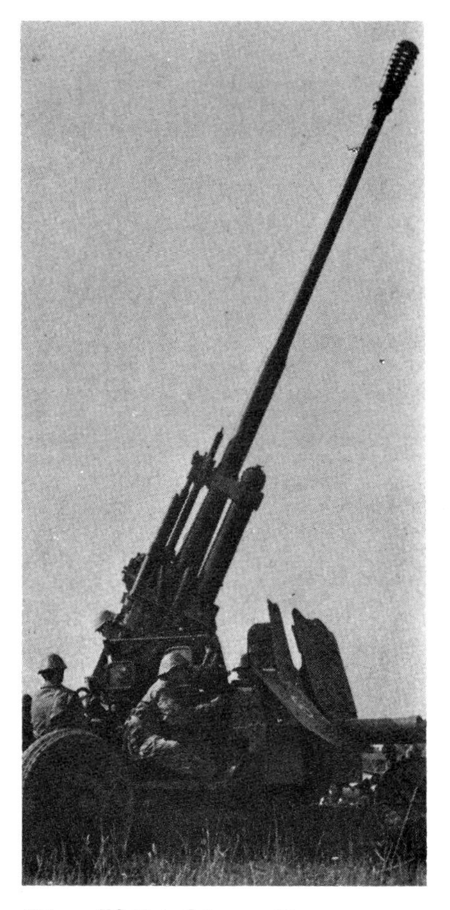

100mm KS-19 in firing position

Employment

Cambodia, China, Cuba, East Germany, Egypt, North Vietnam, Poland, Romania.

M-1939 KS-12 85mm Anti-Aircraft Gun
M-1944 KS-18 85mm Anti-Aircraft Gun
Soviet Union

Calibre: 85mm
Weight Firing: 4300kg
Length Travelling: 7049mm
Width Travelling: 2150mm
Height Travelling: 2250mm
Track: 1800mm
Ground Clearance: 400mm
Barrel Length: 4693mm
Elevation: -3° $+82^\circ$
Traverse: 360°
Maximum Range: 15,650m Horizontal
10,500m Vertical
8380m Vertical (effective)
Rate of Fire: 15–20rpm
Ammunition: HE wt 9·5kg, m/v 792 m/s
APHE wt 9·3kg, m/v 792 m/s
HVAP wt 5·0kg, m/v 1030 m/s
Armour Penetration: APHE 102mm at 1000m
HVAP 130mm at 1000m
Crew: 7
Towing Vehicle: ZIL-157, 6×6, Truck
Above data relates to the KS-12

The KS-12 entered service with the Soviet Army shortly before the start of World War II, the weapon is no longer in service with the Soviet Army. The basic gun was also used later in the war for tank, assault and anti-tank guns.

The weapon has a semi-automatic vertical sliding wedge breech block, a hydropneumatic recuperator and a hydraulic recoil buffer. Sights are provided on the carriage for both anti-aircraft and anti-tank roles. The PUAZO-6/12 director and SON-9 (or SON-9A) fire control radar can be used with the gun. A new fire control radar, the 'FANSONG-E' is replacing the SON-9 and 9A radars.

The KS-18 has greater performance including a higher muzzle velocity. Recognition features are its long barrel with a conical multi-baffle muzzle brake, it can be seen with or without its shield, recoil system above and below the barrel. The gun is mounted on a pedestal mount with a circular platform that is raised when in the travelling position. It has a four-wheel carriage, a swinging arm either side, each with a jack, and a jack at the front and rear of the carriage to support the carriage when in the firing position.

This weapon is also known as the M-1939 and M-1944 Medium Anti-Aircraft Gun.

A modified version of this weapon was built in Czechoslovakia, this had a longer barrel and a 'T' shaped muzzle brake.

Employment

KS-12 China, North Vietnam
KS-18 Albania, Bulgaria, China, Cuba, Czechoslovakia, Cambodia, East Germany, Egypt, Hungary, Iran, North Korea, North Vietnam, Poland, Sudan, Syria, Yugoslavia.

85mm M-1939 KS-12 Anti-Aircraft Gun in action during World War II

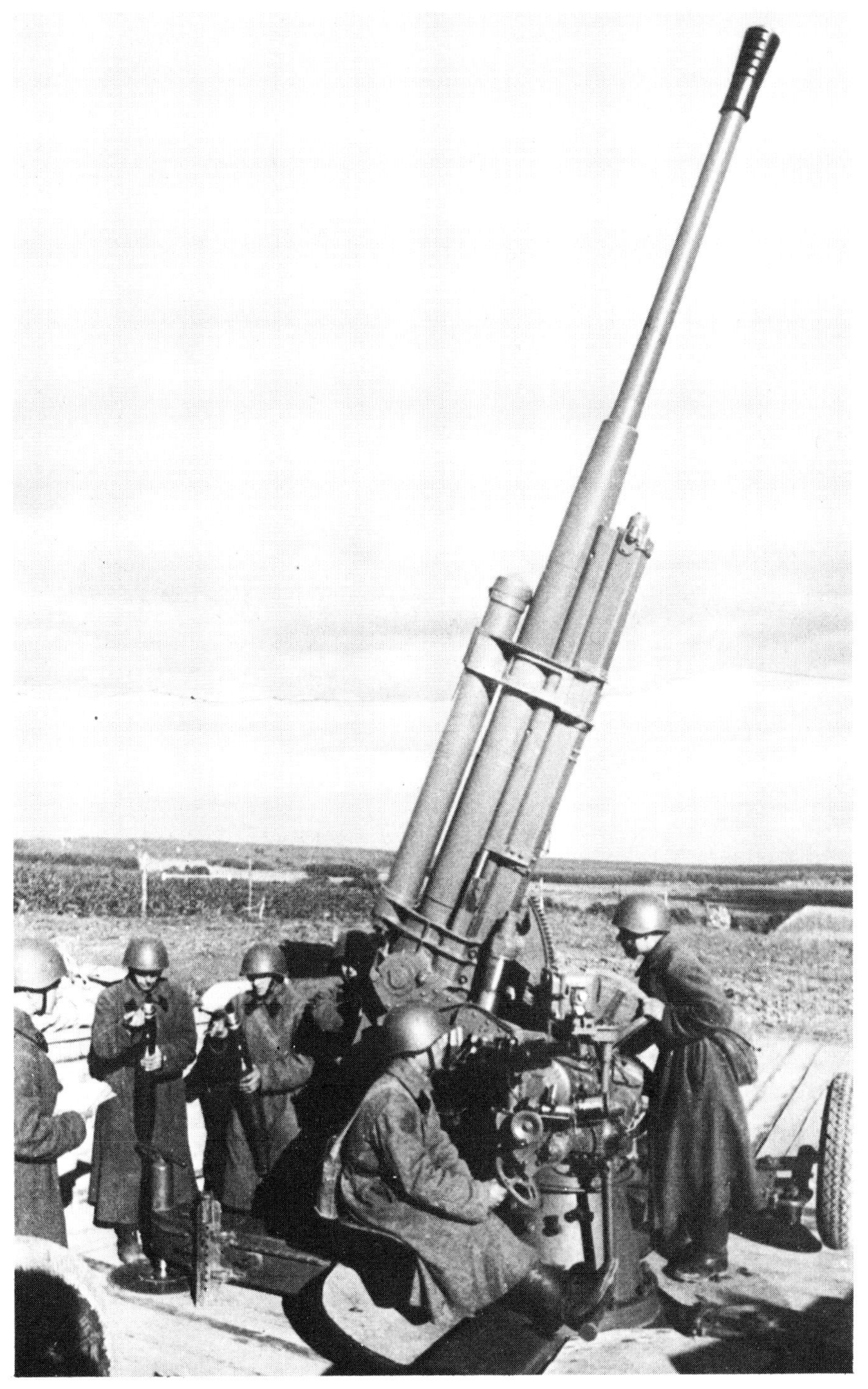

M-1950s in firing position

57mm M-1950 (S-60) Light Anti-Aircraft Gun

Soviet Union

Calibre: 57mm
Weight Firing: 4500kg
Weight Travelling: 4660kg
Length Travelling: 8500mm
Width Travelling: 2054mm
Ground Clearance: 380mm
Barrel Length: 4390mm (including muzzle brake)
Elevation: -4° $+85^\circ$
Traverse: 360°
Maximum Range—Horizontal: 12,000m
Maximum Range—Vertical: 8800m
4000m (effective on carriage)
6000m (effective on jacks)
Rate of Fire: 70rpm (practical)
Ammunition: HE wt 2·8kg, m/v 1000 m/s
APHE wt 3·1kg, m/v 1000 m/s
Armour Penetration: 106mm at 500m with APHE round
Crew: 7
Towing Vehicle: Ural-375, 6×6, Truck
ZIL-157, 6×6, Truck

This versatile anti-aircraft gun was first seen in 1950 and is based on the German Type 58 55mm L/70·7 Anti-Aircraft Gun. The gun is the same as that used in the ZSU-57-2 Self-Propelled A/A weapon as well as the ASU-57 (Self-Propelled) and M-1943 and M-1955 anti-tank guns.

The ammunition is in clips of 4 and fed into the gun from the left, gun operation is recoil and fully automatic. The ammunition has a proximity fuse fitted. A computing sight is mounted on the carriage, or the SON-9 and SON-9A radar and PUAZO 6/60 director used off the carriage. The gun can also be used with a new Fire-Control Radar called 'Fire-Can' which is S-Band.

The weapon is mounted on a four-wheel carriage that has two folding outriggers and four hand-operated levelling jacks. The wheels are not removed when the weapon is fired.

It is recognisable by its four-wheeled carriage with suspension at 45° to the horizontal, long barrel with a pepperpot muzzle brake, barrel support at the rear of the carriage, front of the shield folds down for travelling and there are three seats to the rear of the weapon.

Employment
Albania, Bulgaria, Cambodia, China, Czechoslovakia, East Germany, Egypt, Indonesia, Iran, North Korea, North Vietnam, South Yemen, Poland, Romania, Somalia, Syria, Yugoslavia, Mongolia.

M-1939 37mm Light Anti-Aircraft Gun

Soviet Union

Calibre: 37mm
Weight Firing: 2000kg
Length Travelling: 6160mm
Width Travelling: 1710mm
Height: 2700mm
Track: 1545mm
Ground Clearance: 390mm
Barrel Length: 2575mm
Elevation: -5° $+85^\circ$
Traverse: 360°
Maximum Range: 8000m (Horizontal)
6000m (Vertical)
1400m (Vertical-effective)
Rate of Fire: 80rpm (practical)
Ammunition: HE wt 0·74kg, m/v 880 m/s
AP wt 0·77kg, m/v 880 m/s
HVAP wt 0·62kg, m/v 960 m/s
Armour Penetration: 46mm at 500m (AP round)
Crew: 8

The M-1939 is a modified M-1938 Light Anti-Aircraft Gun and is based on the Swedish Bofors design and is similar to British and American weapons of the same period. It is no longer in service with the Soviet Army having been replaced by the more powerful S-60 57mm weapon.

It has a rising breech system. The ammunition is in clips of five rounds which are fed from the top, there is a curved ejection chute for the empty cases underneath the breech. The recoil system projects a short distance beyond the tube jacket.

The weapon can be seen with or without the shield and is mounted on a four-wheeled carriage, it has been noted that there are two types of disc wheel, one with two holes and the other with five holes. When in the firing position the wheels are removed and the weapon rests on four jacks (one at each end of the carriage, the other two on an outrigger that swings out either side of the carriage), there is also a travelling lock for the barrel at the rear of the carriage.

The Chinese Communists call it the 37mm Anti-Aircraft Gun Type 55. The Algerian Army have a twin version of this weapon in service, although the Algerian weapon has been reported also as the Soviet Naval M-1944 37mm S-60 37mm Twin Light Anti-Aircraft Gun.

Employment
Albania, Bulgaria, China, Cuba, East Germany, Egypt, Mongolia, North Korea, North Vietnam, North Yemen, Romania, Syria, Tanzania, Yugoslavia.

M-1939 37mm Light Anti-Aircraft Gun without its shield

ZU-23 in the firing position

23mm ZU-23 Light Anti-Aircraft Gun (Twin)

Soviet Union

Calibre: 23mm
Weight Firing: 1000kg
Length Travelling: 4648mm
Width Travelling: 2774mm
Height Travelling: 2073mm
Track: 1670mm
Barrel Length: 2010mm
Elevation: −10° +90°
Traverse: 360°
Maximum Range—Horizontal: 7000m (2500m practical)
Maximum Range—Vertical: 5000m (1500m practical)
Rate of Fire: 200rpm (per barrel practical)
Ammunition: HEI wt 0·19kg, m/v 970 m/s API wt 0·19kg, m/v 970 m/s
Armour Penetration: 25mm at 500m (API round)
Crew: 5
Towing Vehicle: GAZ-69, 4×4, Truck

The ZU-23 has replaced the ZPU-4 in many units of the Warsaw Pact Forces. A quad version of this weapon is mounted on a tracked chassis and called the ZSU-23-4.

The weapon has quick change barrels, cyclic rate of fire is 1000 rounds per minute, per barrel. Operation is gas, fully automatic, and has a vertical sliding wedge type breechblock. The twin barrels have flash eliminators fitted. The ammunition is similar to that used in the VYa aircraft cannon, the ammunition boxes are each side of each barrel and each contain 50 rounds of belted ammunition.

Recognition features are its two-wheeled carriage with a small jack at the front and rear and the cylindrical flash eliminators. The wheels go almost flat on the ground when the weapon is in the firing position.

Employment
Czechoslovakia, East Germany, Egypt, Finland, North Vietnam, Poland, Soviet Union, Syria.

BELOW
Twin 14·5mm Anti-Aircraft MG ZPU-2 (new model). Note the ammunition boxes.

FOOT
Twin 23mm Automatic Anti-Aircraft Gun ZU-23. Compare this with the above weapon.

ZPU-1 14·5mm Heavy Anti-Aircraft Machine Gun

Soviet Union

Calibre: 14·5mm
Weight Firing: 581kg
Length Travelling: 3871mm
Width Travelling: 1372mm
Height Travelling: 1100mm
Barrel Length: 1359mm
Elevation: −15° +85°
Traverse: 360°
Maximum Range—Horizontal: 7000m
Maximum Range—Vertical: 5000m
1400m (effective vertical range)
Ammunition: API wt 64·4gr, m/v 1000 m/s
Armour Penetration: 32mm at 500m (API round)
Crew: 3
Towing Vehicle: GAZ-69, 4×4, Truck

This is a single version of the Vladimirov (KPV) heavy machine gun, mounted on a two-wheeled, single-axle carriage. It has quick change barrels and the magazine holds 150 rounds of belted ammunition, this is on the right side. For the A/A role it has a reflex optical sight and for the ground role a telescope. It is no longer in service with any of the Warsaw Pact Forces.

Recognition features are its single barrel with a flash eliminator, two-wheeled carriage similar to the Model (late) 2 ZPU-2, when in firing position the wheels are raised.

Employment
North Vietnam, United Arab Republic.

ZPU-2 14·5mm Twin Heavy Anti-Aircraft Machine Gun

Soviet Union

Calibre: 14·5mm
Weight Firing: 635kg
Length Travelling: 3871mm
Width Travelling: 1372mm
Height Travelling: 1100mm
Track: 1100mm
Ground Clearance: 270mm
Barrel Length: 1359mm
Elevation: −15° +85°
Traverse: 360°
Maximum Range—Horizontal: 7000m
Maximum Range—Vertical: 5000m
1400m (effective vertical range)
Rate of Fire (per barrel): 150/200rpm
Cyclic Rate of Fire (per barrel): 600rpm
Ammunition: API wt 64·4gr, m/v 1000 m/s
Armour Penetration: 32mm at 500m (API round)
Crew: 4
Towing Vehicle: GAZ-69, 4×4, Truck

This is a twin version of the Vladimirov (KPV) Heavy Machine Gun. There are two models:

Model 1—Original model with two large mudguards, double tubular towbar, wheels removed for firing and weapon rests on 3 points.

Model 2—Later model with two small mudguards, lightweight single towbar and lower silhouette. The wheels are raised when in the firing position.

System of operation is recoil, fully automatic. The ammunition is in metal linked belts of 150 rounds, each barrel has one magazine of 150 rounds. Ammunition is the same as that used in the obsolete PTRS (Simonov) and PTRD (Degtjarev) anti-tank rifles.

Its recognition features are its two-wheeled carriage, one large ammunition box either side, holes for cooling the barrel, flash eliminators, and sighting mechanism high above the barrels.

These weapons have also been seen to be mounted on the K-61 tracked amphibian, BTR-40 and BTR-152 Armoured Personnel Carriers.

Employment
China, Cuba, East Germany, Egypt, Hungary, Poland, Romania, Bulgaria, Soviet Union, Syria, North Korea, North Vietnam.

ZPU-4 14·5mm Quad Anti-Aircraft Machine Gun

Soviet Union

Calibre: 14·5mm
Weight Firing: 2000kg
Length Travelling: 4979mm
Width Travelling: 1676mm
Height Travelling: 2103mm
Ground Clearance: 458mm
Barrel Length: 1359mm
Elevation: -10° $+90^\circ$
Traverse: 360°
Maximum Range—Horizontal: 7000m
Maximum Range—Vertical: 5000m
2000m (effective horizontal range)
1400m (effective vertical range)
Rate of Fire Per Barrel: 150rpm
Cyclic Rate of Fire Per Barrel: 600rpm
Ammunition: API wt 64·4gr, m/v 1000 m/s
also IT and APIT
Armour Penetration: 32mm at 500m (API round)
Crew: 5
Towing Vehicle: GAZ-63, 4×4, Truck
GAZ-54, 4×4 Truck

This weapon is a quad version of the KPV 14·5mm Heavy Machine Gun. Each barrel has 150 rounds of ammunition on a metal link belt in a drum type magazine. The barrels can be quickly changed. System of operation is recoil, full automatic and air cooled.

For ground targets a telescopic sight is provided and for the anti-aircraft role a reflector sight is provided.

The guns are mounted on a four-wheeled carriage with a travelling lock for the guns at the rear, the sight is the highest part of the weapon. When firing the weapon is supported by the 4 stabilizing jacks, screw jack at the front and rear and swinging arms with a jack either side of the carriage. The weapon can be fired with the wheels on the ground.

Recognition features are its four-wheeled carriage, 4 barrels with drum type ammunition containers and no shield.

Employment
Egypt, North Vietnam.

ZPU-4 in travelling position

75mm Anti-Tank Gun — Spain

Calibre: 75mm
Weight: 1425kg
Length Overall: 5840mm
Barrel Length: 3450mm
Elevation: -5° $+22^\circ$
Traverse: 65° (total)
Range: 7700m (HE round)
Ammunition: HEAT wt 7·8kg, m/v450 m/s
HE wt 9·1kg, m/v 550 m/s
APC wt 11·9kg, m/v 770 m/s

This is used in some numbers as an anti-tank gun by infantry regiments.

105mm/46 Howitzer — Spain

Calibre: 105mm
Weight: 1950kg
Length Firing: 5860mm
Length Travelling: 6080mm
Width: 2100mm
Height (Barrel elevated): 2950mm
Track: 1732mm
Barrel Length: 3350mm
Elevation: -5° to $+45^\circ$
Traverse: 50° (total)
Range: 11,450m
8500m (effective)
Rate of Fire: 4 rounds in first ½ minute
16 rounds in first 4 minutes
Ammunition: Spanish round wt 15·27kg, m/v 504 m/s
American round wt 14·97kg, m/v 476 m/s

This was developed in the early 1950s from an earlier 1943 weapon, the latter is still used as a training weapon as well as a saluting piece.

122mm/46 — Spain

Calibre: 122mm
Weight Firing: 7100kg
Weight Travelling: 7800kg
Length Firing: 8800mm
Length Travelling: 9000mm
Width: 2350mm
Elevation: -4° to $+45^\circ$ (Model 390/1)
-2° to $+65^\circ$ (Model 390/2)
Traverse: 56° (total)
Range: 20,000m (maximum)
Ammunition: HE wt 25kg, m/v 550–830 m/s

This was originally a Soviet weapon captured by the Germans during World War II and transferred to Spain, some were also built in Spain. A few of these are in service and some are held in reserve.

ECIA 81mm Model L and L1 Mortars

There are two models both with a calibre of 81·35mm, they are: *Model L* barrel length 1150mm, maximum range 4100m, weight firing 43kg, rate of fire 15rpm. *Model L1* barrel length 1450mm, maximum range 4500m, weight firing 45kg, rate of fire 15rpm.

ECIA 120mm Model SL Mortar

Basic data is as follows: calibre 120mm, barrel length 1600mm, maximum range 5940m, weight firing 123kg, weight travelling 257kg, rate of fire 12rpm.

ECIA 120mm Model L Mortar

Basic data is as follows: calibre 120mm, barrel length 1600mm, maximum range 6660m, weight firing 213kg, weight travelling 361kg.

ECIA 105mm Model L Mortar

Basic data is as follows: calibre 105mm, barrel length 1500mm, maximum range 7050m, weight firing 105kg, weight travelling 239kg, rate of fire 12rpm.

Earlier mortars in use are the 81mm (1942) and 120mm (1942), these are still used for training purposes.

Employment

All of the above are used by the Spanish Army. The Spanish Marines use 81mm 1951 mortar. Spain is also reported to have a large amount of coastal artillery including 381mm (15in) and 152mm (6in) dating from the 1920s.

RIGHT
ECIA 81mm Mortar, as used by the Spanish Army and Marines

BELOW
120mm Heavy Mortar being towed by a M-113 of the Spanish Army

FOOT
105mm/26 towed Howitzers of the Spanish Army, trucks are M-34s

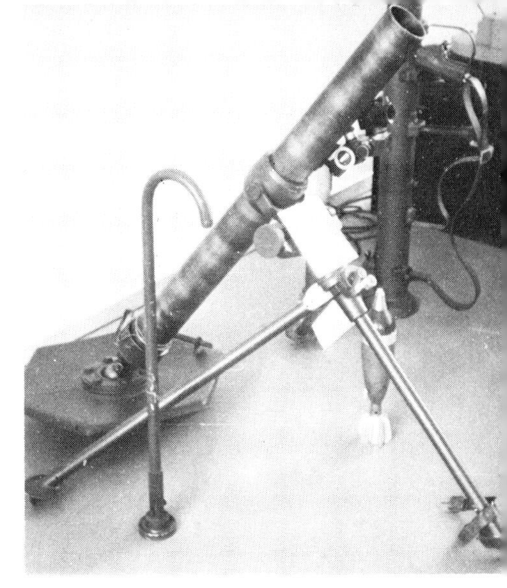

155mm Model 'F' Field Howitzer — Sweden

Calibre: 155mm
Weight Travelling: 9000kg
Length Firing: 7150mm
Length Travelling: 7800mm
Width Firing: 6800mm
Width Travelling: 2750mm
Height Firing: 1650mm
Height Travelling: 2500mm
Elevation: −4° +69°
Traverse: 40° left and right
Range: 17,600m
Rate of Fire: 3–4rpm
Ammunition: Projectile wt 43kg, m/v 650 m/s
Towing Vehicle: Volvo TL31 L3153, 6×6, Truck

This is the French 155mm M-1950 weapon built in Sweden. It is provided with a split trail with heavy spades on each end, a castor wheel is fitted to the right trail to assist in preparing the weapon for firing. There is a firing jack at the front of the carriage. It is reported that this is a difficult weapon to operate and maintain. The barrel has a multi-baffle muzzle brake. Also refer to the French section.

Employment
Sweden.

150mm m/39 Field Howitzer — Sweden

Calibre: 149·1mm
Weight: 5720kg
Length Firing: 7270mm
Length Travelling (barrel to rear): 6550mm
Length Travelling (barrel forward): 7370mm
Width Firing: 5330mm
Width Travelling: 2500mm
Height Travelling: 2500mm
Barrel Length: 3600mm
Elevation: −5° +66°
Traverse: 45° (total)
Range: 14,600m
Ammunition: Projectile wt 41·5kg, m/v 580 m/s, 8 charges
Rate of Fire: 4–6rpm
Towing Vehicle: Volvo TL31 L3153, 6×6, Truck

Whilst in the travelling position the barrel is drawn to the rear, and chain drive being provided on the right hand side of the carriage for this purpose. The trails are of the split type, each trail being provided with a spade. There is also a castor wheel to assist in opening out the trails. The m/39 is fitted with old type wheels with solid rubber tyres whilst the m/39b has modern type wheels and rubber tyres.

When in action the weapon is very similar in appearance to the British 5·5in gun.

Employment
Sweden.

TOP
Two photographs show the 155mm Model 'F' weapon in firing position

RIGHT
The m/39b. The barrel in this photograph has been drawn to the rear. The spades can be seen on the top of the trails, as is the castor wheel.

105mm 4140 Field Howitzer in firing position. The central firing jack can be seen just under the wheel.

105mm 4140 Field Howitzer — Sweden

Calibre: 105mm
Weight Firing: 2800kg
Weight Travelling: 3000kg
Length Firing: 5520mm
Length Travelling: 6800mm
Width Firing: 4900mm
Width Travelling: 1810mm
Height Firing: 1200mm
Height Travelling: 1850mm
Barrel Length: 2940mm (28 calibres)
Elevation: −5° +65°
Traverse: 360°
Range: 14,600m
Rate of Fire: 25rpm (maximum)
8rpm (normal)
Ammunition: HE, projectile weight 15·5kg
m/v 607 m/s, 8 charges
Towing Vehicle: Volvo TL31 L3153, 6×6, Truck

The 4140 Field Howitzer is built in Sweden by Bofors. It is provided with a small shield, the barrel has a single stage baffle brake. When in the firing position it rests on 4 trails, each of the trails being provided with a picket. There is also a central firing jack. It is easily recognisable by its small square shield very much to the rear and when in the firing position the wheels are off the ground and in front of the shield.

Employment
Sweden.

NOTE: *Switzerland has purchased a number of Bofors weapons and these have the following Swiss designations 105mm Howitzer 46,105mm Gun 35 and 150mm Howitzer 42, the latter is reported to be the m/39.*

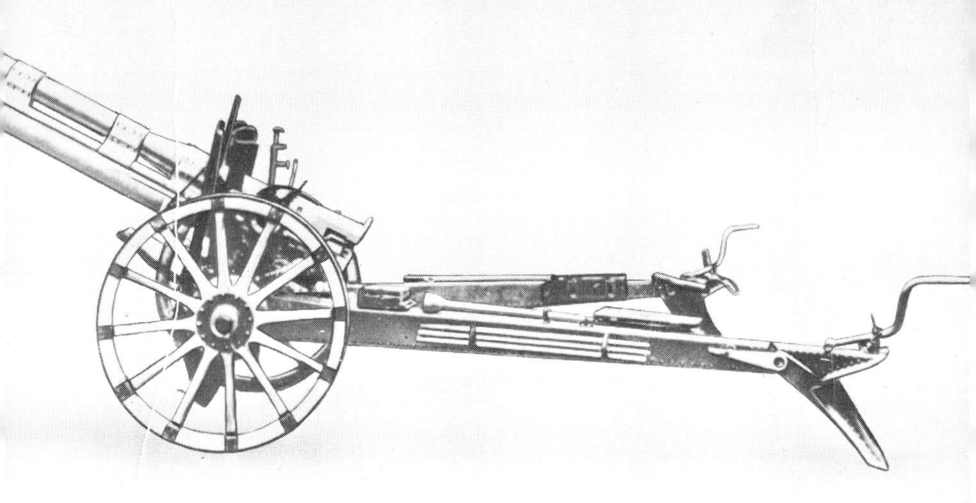

105mm m/40 Howitzer

105mm m/40 Field Howitzer — Sweden

Calibre: 105mm
Weight Firing: 1850kg
Weight Travelling: 1900kg
Length Firing: 5500mm
Length Travelling: 6050mm
Width Firing: 4000mm
Width Travelling: 1900mm
Height Firing: 1130mm
Height Travelling: 1700mm
Elevation: -5° $+45^\circ$
Traverse: 50° (total)
Range: 14,300m (charge 8)
13,000m (charge 7)
Rate of Fire: 10rpm
Ammunition: m/34 wt 15·5kg, m/v 560 m/s, 7 charges, HE
m/60 wt 14·5kg, m/v 620 m/s, 8 charges, HE
Towing Vehicle: Tractor

Mounted on a carriage with split trails, each of the trails has a spade at each end. The wheels are of the old spoked type. A shield is provided, this extends to below the axles. Very short barrel with no muzzle device.

Employment
Sweden.

75mm M-1902 Howitzer — Sweden

This very old Mountain Pack Howitzer is reported to be still used by the Turkish Army.

90mm PV-1110 Recoilless Rifle in firing position

90mm PV-1110 Recoilless Rifle — Sweden

Calibre: 90mm
Total Weight Firing & Travelling: 260kg
Weight of Gun: 125kg
Weight of Carriage: 135kg
Length Firing: 3691mm
Length Travelling: 4100mm
Width Travelling: 1375mm
Height Travelling: 870mm
Barrel Length: 3691mm
Elevation: −10° +15°
Traverse: 75° to 110° (total) depending on the firing height
Range: 700m (effective)
Rate of Fire: 2 rounds in 13 seconds
6 rounds in 1 minute
Ammunition: Complete round weighs 9·6kg, projectile weight 3·1kg, m/v 715 m/s, fin-stabilised, hollow charge
Armour Penetration: 380mm at 90°, range 700m
Crew: 2–3
Towing Vehicle: Volvo, 4×4, Laplander Haflinger 4×4

The 90mm PV-1110 (PV standing for pvpjäs) is built by Bofors in Sweden. An interesting feature of the weapon is that its overall height can be adjusted from 870mm to 470mm. The weapon is aimed by a 7·62mm spotting rifle (10 rounds for ready use) mounted over the barrel, the sight is to the left of the barrel, the weapon is fired with the aid of a pistol type grip under the barrel.

The weapon can also be mounted on a sled, a Volvo, 4×4, Laplander or an Austrian Haflinger vehicle, it can also be towed by the latter vehicles. When used in the Swedish infantry brigades it is normally towed and when with the Swedish armoured brigades it is mounted on the Volvo Laplander vehicle.

Employment
Sweden.

TOP
90mm PV-1110 Recoilless Rifle mounted on a sled

ABOVE
90mm PV-1110 Recoilless Rifle being towed by a Haflinger 4×4 vehicle

The m/54 57mm Anti-Aircraft Gun in firing position

57mm m/54 Anti-Aircraft Gun — Sweden

Calibre: 57mm
Weight: 8100kg
Length Firing: 7270mm
Length Travelling: 8435mm
Width Firing: 3800mm
Width Travelling: 2380mm
Height Firing: 2435mm
Height Travelling: 2810mm
Barrel Length: 3420mm
Elevation: -5° $+90^\circ$
Traverse: 360°
Rate of Fire: 160rpm (cyclic)
Ammunition: Weight 2·6kg, m/v 920 m/s, time to 3000m=4·2 seconds, time to 5000m =8·5 seconds
Range: Effective A/A is 4000m
Max ground range 14,000m
Towing Vehicle: Volvo TL31 L3154, 6×6, Truck

This was built in Sweden by Bofors. It is mounted on a four-wheeled carriage, when in the firing position the wheels are kept on and the carriage rests on 4 jacks, one at each end of the carriage and another either side mounted on stabilisers that swing out from the side of the carriage.

The weapon is easily recognisable by its very large shield which is flat at the front and has flat side pieces to it. Ready use ammunition racks are provided at the rear of the platform. As with all of the Bofors A/A guns the barrel is provided with a flash hider. This weapon can also be used in conjunction with a Fire Control System, i.e. similar to the 40mm L/70.

Employment
Belgium, Sweden.

40mm Light Anti-Aircraft Gun m/48 (L/70)
Sweden

Calibre: 40mm
Weight: 4800kg
Length Overall Travelling: 7290mm
Length Excluding Drawbar: 6270mm
Width Firing: 4440mm
Width Travelling: 2250mm
Height Firing: 1940mm
Height Travelling: 2350mm
Barrel Length: 2800mm
Elevation: -5° $+90^\circ$
Traverse: 360°
Rate of Fire: 240rpm (cyclic)
Ammunition: Fixed—HE, AP-T, APDS-T. Weight 0·96kg, m/v 850/1000 m/s
Range: 3000m effective anti-aircraft range
Crew: 6, of which 2 are on gun, the other 4 handling or loading ammunition
Towing Vehicle: Bedford, 4×4, Truck (UK)
AEC, 6×6, 10 Ton Truck (UK)
Volvo TL22 L2204, 6×6, Truck (Sweden)

This gun was first developed after the end of World War II, the first prototype being built in 1947. In 1951 it entered production and since then it has been exported all over the world both for land use and for use in warships.

The carriage has racks for a total of 48 rounds of ammunition, the ammunition is fed into guideways above the breech, a total of 16 rounds can be held above the breech for immediate use. The empty cartridge cases are ejected via a chute in the lower half of the shield. The weapon is mounted on a four-wheel carriage. When in firing position the weapon rests on 4 jacks, one at each end ot the carriage and the other 2 on stabilisers that swing out either side of the carriage.

There are four models—m/48, m/48D, m/48R and m/48C, the differences being in their carriages. Some of the models have an integral power unit on the carriage. The m/48D and m/48R can be radar controlled. They can be controlled by various types of radar. For example the British Army uses the British developed and built Fire Control Equipment No 7 (Yellow Fever), when using this the system consists of the gun, FCS and a mobile generator. Other countries use the Super-Fledermaus FCS (Swiss built), French Eldorado-Mirador, Dutch L4/5 (built by Hollandse Signaalapparaten) and the Swedish Army uses the PS-04/R. The L/70 has also been built in many other countries including Italy.

Employment
Austria (FIAK-57), France, India, Israel, Italy, Spain, Sweden, Great Britain.

Spanish 40mm Anti-Aircraft Gun, with built in generator

TOP
The British Fire Control System used in conjunction with the British Army's 40mm Anti-Aircraft Guns

ABOVE
Swedish 40mm Anti-Aircraft Gun. Note the ready use ammunition in an arc above the breech.

BELOW
Swedish 40mm Anti-Aircraft Gun. The ejection chute for the empty cartridge cases can be seen at the bottom of the shield below the barrel.

40mm m/36 Light Anti-Aircraft Gun Sweden

Calibre: 40mm
Weight: 2150kg (m/38)
2400kg (m/39)
2050kg (m/48E)
Length Travelling: 6300mm (m/38)
6380mm (m/39)
6390mm (m/48E)
Width Firing: 3980mm (m/38)
3920mm (m/39)
3920mm (m/48E)
Width Travelling: 1720mm (all models)
Barrel Length: 2240mm (56 cal)
Elevation: $-5°$ $+90°$
Traverse: 360°
Rate of Fire: 120/140rpm (cyclic)
70rpm (practical)
Ammunition: 800/880 m/s
Range: 3000m effective A/A

This is the earliest of the Bofors anti-aircraft guns which was to become famous during World War II. It is mounted on a four-wheel carriage with no shield, when in firing position the weapon rests on jacks at either end of the carriage and stabilisers on outriggers either side of the carriage. There were three types of carriage—m/38, m/39 and m/48E.

During World War II it was also built in the United States, Great Britain and Canada. Before the war it was built in Hungary and Czechoslovakia.

Employment
Austria, Burma, Cambodia, Cyprus, Egypt, Eire, Finland, India, Israel, Ivory Coast, Jordan, Lebanon, Libya, Malaysia, Nigeria, Sudan, Tunisia, Yugoslavia and Zaire.

Swedish 40mm m/36 Light Anti-Aircraft Gun

120mm m/41C (1941) Mortar — Sweden

Calibre: 120·25mm
Weight Firing: 285kg
Weight Travelling: 600kg
Barrel Length: 2000mm
Elevation: +45° +80°
Traverse: 360°
Rate of Fire: 12–15rpm
Range: 6400m
Ammunition: HE shell weight 13·3kg, m/v 125/317 m/s, plus other types of bomb

This is the Finnish Tampella M-1940 120mm Mortar produced in Sweden. In 1956 the M-56 stand (produced under licence from Hotchkiss Brandt of France) replaced the earlier Tampella-type stand. 8 m/41Cs are in the mortar company of the infantry brigades and 6 m/41Cs in the mortar company of the Norrlands Brigades. In 1972 the m/41Cs were fitted with new sights.

Employment
Eire and Sweden.

120mm m/41C Mortar in firing position

81mm Mortar m/29 (1929) — Sweden

Calibre: 81·4mm
Weight Firing: 60kg
Barrel Length: 1000mm
Elevation: +45° +80°
Traverse: 90° (total)
Rate of Fire: 15–18rpm
Maximum Range: 2600m
Ammunition: HE shell weight 3·5kg, m/v 70/190 m/s
Crew: 2–3

This is the French Stokes-Brandt 81mm M-1917 Mortar produced in Sweden, it was built until 1943. In 1934 Swedish sights replaced the French sights. 6 m/29s are in the mortar platoon of the heavy weapons company of each Swedish infantry battalion.

The m/29 is very similar in appearance to the American 81mm mortar M-1, they both have a rectangular base with a carrying handle, bipod with chains between them and the left leg has an anti-cant device fitted to it.

Employment
Sweden, Eire.

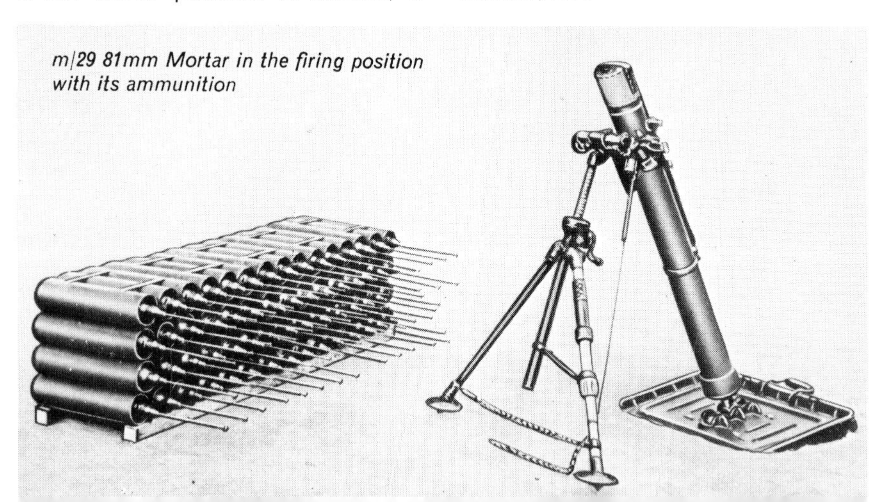

m/29 81mm Mortar in the firing position with its ammunition

20mm HO-639-B Gun with ammunition box containing 75 rounds of ammunition

Hispano-Suiza Anti-Aircraft Guns

Switzerland

NOTE: *In 1971 Oerlikon-Bührle took over the military division of Hispano-Suiza. This has changed the designations of the guns, i.e. the HSS-665 is now the Ho-665.*

HSS-820SL 20mm gun

This can have four types of ammunition feed: (1) A 50 round magazine. (2) Open-link belt feed from a box-magazine with left-hand feed. (3) As No 2 but with right hand feed. (4) A vertical box magazine above the weapon. Rate of fire is adjustable between 800 and 1060rpm, muzzle velocity is 1050 m/s. Weight of the gun without the magazine is 57kg. The following types of mount are used:

HSS-639 (single 20mm)

This has a Delta-4 optical reflex sight with ×2·3 or ×4 magnification. It can be used against both ground and air targets and has a traverse of 360°, elevation −7° to +83° by handwheel. It is mounted on a tripod type base. Its travelling weight is 512kg. The HSS-639-B1 has a 50 round drum magazine and the HSS-639-B3 has a 75 round belt. The HSS-639 is known in the United States as the M-139.

HSS-665 (triple 20mm)

This has a Galileo P36 tracking and control unit for air and land targets. Traverse is 360°, elevation −5° to +83°. Has a tripod type base. Each barrel is provided with a drum type magazine holding 50 rounds. Its travelling weight is 1230kg.

HSS-666 (twin 20mm)

This has the Galileo control unit P36, travelling weight is 1200kg. It has two ammunition boxes each holding 120 rounds of ammunition.

HSS-669 (single 20mm)

This has a HSS Kern-722 sight with a ×5 magnification, traverse is 360°, elevation is −10° to +80°. The backrest and seat can be quickly removed allowing the gun to be used against ground targets.

HSS-673 (single 20mm)

This is a very light infantry mount, the

two wheels are detachable and the trails form the basis of the mount. The gunner lays prone behind the gun. In this position traverse is 25½°, elevation −10° to +25°. The ammunition is fed from a four 10-round box magazines carried on the mount. It can be quickly dismantled into 8 manpack loads, total weight (i.e. the gun, mount and trails) is 217kg and the four magazines weigh 44kg.

HSS-676
This is for fitting to APCs.

HSS-801
A triple 20mm Light Anti-Aircraft Gun.

HSS-804
This is a single 20mm Light Anti-Aircraft Gun but has been produced in Yugoslavia as the M-55, in this case it is a triple gun.

30mm HSS-831SL Switzerland

This has four types of feed:

1 An offset magazine with 8×5 round clips, fed from the left.
2 An open link belt fed from the left.
3 As above but from the right.
4 Offset magazine with 8×5 round clips, fed from the right.

Rate of fire is 650rpm and muzzle velocity is 1080 m/s. Weight without magazine is 180kg. Versions include:

HSS-661
This has a Galileo P36 sight and can be used both in the anti-aircraft and ground roles. Traverse is 360°, elevation −5° to +83°. Its control mechanism consists of four main parts—joy stick and tracking aid, a hydraulic power unit, a computer sight (the P36) and a manually operated laying unit. A small Wankel engine supplies the power. The two wheels can be detached to allow use of the tripod-type base. Rate of fire is 650rpm, weight firing is 1150kg, weight travelling 1550kg and m/v 1080 m/s. Dimensions are: barrel length 3·4m, length towed 5·23m, height towed 1·97m and carriage length 1·92m.

The HSS-831SL is also used in many self-propelled anti-aircraft weapons system, e.g. the AMX-30 anti-aircraft tank.

HSS K43 and K58 (20mm guns) Switzerland

This is a 20mm Light Anti-Aircraft Gun mounted on a four-wheeled trailer.

HSS-83/84 (30mm guns) Switzerland

This is a single 30mm Light Anti-Aircraft Gun.

Employment

HS-804 Yugoslavia (built as the M-57) and Sweden
HS-820SL Belgium, India, Italy, Jordan, Netherlands, Pakistan, United States, West Germany.
HS-83/84 Egypt, Finland (ITK-30), Israel, Italy.

TOP RIGHT
20mm HO-665 Triple Gun with three drum type magazines each containing 50 rounds of ammunition

CENTRE RIGHT
20mm HO-666 twin guns, each box magazine holds 120 rounds of ammunition

RIGHT
30mm HO-661 Gun

Oerlikon Guns

Switzerland

Type 204 GK

Calibre: 20mm
Length of Gun in Calibres: 85
Weight of Barrel and Muzzle Brake: 26·6kg
Weight of Complete Gun: 87·0kg
Rate of Fire: 1000rpm

Ammunition:

Type	*Cartridge wt*	*Projectile wt*	*M/V M/S*
Practice SU	340gr	125gr	1100
*Practice SUL**	340gr	125gr	1100
HE SSB/K	338gr	122gr	1100
*Fragment SSBL/K**	340gr	125·4gr	1100
HE MSB/K	320gr	102gr	1200
*HE MSBL/K**	332gr	114gr	1150
APHE PSBH/B	344gr	128gr	1085
APHE PSBLH/B	344gr	128gr	1085
*AP PKLH**	332gr	114gr	1150

* with tracer

The Model 204 GK 20mm gun was designed for belt feed operation. It is self-contained to the extent that the cocking mechanism and ammunition feed mechanism are built integrally with the gun. The ammunition feed mechanism is gas-operated, and because of this the recoil mechanism requires no critical adjustments. In automatic fire the gun is fired during counter-recoil so that it floats in the cradle. This action reduces the recoil forces and eliminates counter-recoil forces on the mount.

The gun is fired from an open bolt and there is little danger of 'cock-offs', the bolt is securely locked when the cartridge is fired. Firing is mechanical. The ammunition feed can be changed from right to left by changing only one part. The barrel and receiver recoil in the cradle. The cover group, with trigger mechanism and belt feed mechanism, is locked to the cradle. A lever on the trigger housing selects for either single round or automatic fire, the other lever on top of the trigger housing adjusts to safe or fire positions.

The weapon is built by Oerlikon and can be fitted to a variety of mounts including: **10 ILa/5TG.** This has a travelling weight of 544kg and a weight in action of 383kg. Its drum type magazine holds 50 rounds. **2 RLa/204G.** Is the 204 GK gun for fitting

20mm Oerlikon Gun as used by Spanish Army

to AFVs. A naval version is called GAM/204GK. There is also a ring mount 1RLA for vehicles.

25mm GB1

This is one of the more recent weapons and can be used for both anti-aircraft and ground fire, or fitted in a turret. It has a calibre of 25mm and fires either HEI or APT ammunition, its two ammunition boxes each hold 40 rounds of ammunition. The gun used is the KBA and has a maximum range of 2000m, m/v of 1110 m/s and a rate of fire of 150/570rpm. Laying of the gun in elevation is done by a handwheel in which the firing trigger is built in. Single shots or bursts can be fired.

20mm K-54

This is a light anti-aircraft gun with a traverse of 360°, elevation −5° to +85°, weight in firing position is 335kg. Effective range in the anti-aircraft role is about 1500m. Rate of fire (cyclic) is about 1000rpm and m/v is 1110 m/s. The gun used is the 204 GK.

ABOVE
20mm Oerlikon Gun type 10 ILa/5TG

BELOW
The new Oerlikon GBI 25mm Gun. Each ammunition box contains 40 rounds each.

K-38 34mm

This a light anti-aircraft gun mounted on a four-wheeled trailer. Its weight in firing position is 280kg and its effective anti-aircraft range is 3000m. Rate of fire is 250rpm and m/v is 900 m/s. There was also a 75mm anti-aircraft gun called the K-38, this is no longer in service.

Employment

Oerlikon 20mm guns are used by many countries including Austria (known as the M-58), Nigeria, Spain, Switzerland, South Africa and Egypt.

Twin 35mm Anti-Aircraft Gun K-63

Switzerland

Calibre: 35mm
Weight (Firing Position): 6400kg
Elevation: −5° +92°
Traverse: 360°
Effective Tactical Range: 4000m
Rate of Fire: 550rpm per barrel cyclic
Ammunition: Includes HE and AP, weight 0·55kg, m/v 1175 m/s
Armour Penetration: 44mm at 1000m with AP round

This weapon is built in Switzerland by Oerlikon, its company designation is 2ZLa/353MK whilst its Swiss designation is K-63 (Flab Kan 63).

It is mounted on a four-wheeled carriage, when in the firing position the wheels are lifted clear of the ground and the weapon rests on four stabilisers, one at each end of the carriage, and one either side of the carriage, the latter being on outriggers.

Elevation speed is 56° a second and traverse speed is 112° a second. The ammunition is fed to the guns on a clip feed conveyor system.

In Switzerland, indeed in many other countries as well, the system is used in conjunction with the Contraves 'Super-Fledermaus' Fire Control System. The weapon has also been produced under licence in Japan. A navalised version of the 2ZLa/353MK system is called the GDM-A.

Employment
Austria (known as the M-65), Finland, Japan, South Africa and Switzerland.

This shows the K-63 (2ZLa/353MK) in firing position. The switching over from travelling position to the firing position is achieved by hydraulic means. The devices on the muzzles of the guns are for measuring the muzzle velocity of the projectiles as they leave the barrel.

8in Howitzer in service with the Spanish Army

8in Howitzer M-115

United States

Calibre: 203·2mm
Weight Firing: 13,471kg
Weight Travelling: 14,515kg
Length Overall: 10,972mm
Width Travelling: 2844mm
Width Firing: 6857mm
Height Travelling: 2743mm
Height Firing: 2133mm
Elevation: $-2°$ $+65°$
Traverse: 30° left and right
Rate of Fire: 30 rounds per hour
Range: 16,800m
Ammunition: AE (Atomic Explosive) M-422
HE M-423
HES (HE-Spotting) M-424
HE M-106
Dummy M-14
Gas M-426
RAP M-509
Crew: 14
Emplacement Time: 20 minutes (minimum)
Towing Vehicle: M-125, 10 ton, 6×6, Truck
M-6 Tracked Vehicle

This weapon was developed during World War II and built by the Hughes Tool Company. It was first known as the 8in Howitzer M-1 and Carriage M-1. The M-115 consists of the M-2 Cannon, Recoil Mechanism M-4/M-4A3 on a M-1 Carriage, and the M-5 limber. Some M-115s have the M-2A1 Cannon which is built of stronger metal.

The M-115 has a stepped thread/interrupted screw breechblock, a hydro-pneumatic variable recoil mechanism and a pneumatic equilibrator.

For travelling the trails are locked together and the two-wheel limber attached. For firing the limber is detached and the weapon supported on the twin four-wheel axles of the carriage and the trials at the rear. Two spades are carried on either side of each trail and these are placed on the end of each trail when in firing position. Some armies tow the howitzer without the limber.

This weapon is also mounted on a tracked chassis, system is then known as the M-110. A wartime model was the M-43 but this was built in very small numbers.

Employment
Belgium, Denmark, Germany, Greece, India, Italy, Jordan, Netherlands, Spain, Turkey and Great Britain. USA (Reserve forces).

TOP
Close up of an 8in Howitzer

ABOVE
8in Howitzer in travelling position (photograph taken in Japan)

155mm Gun M-1A1

United States

Calibre: 155mm
Weight Travelling: 13,880kg
Weight Firing: 12,600kg
Length Travelling: 10,434mm
Width Travelling: 3020mm
Height Travelling: 2590mm
Elevation: $-2°$ $+63°$
Traverse: 30° (left and right)
Rate of Fire: 1rpm
Range: 23,500m
Ammunition: HE M-101 wt 57·59kg
AP M-112 wt 59·56kg
m/v 835 m/s
Crew: 14
Towing Vehicle: 10 ton, 6×6, Truck
M-4 Tractor

This was developed from the French 155mm gun, it was standardised in 1938 and was in production by the time World War II started. The gun was known as the 'Long Tom'. The first weapons were the Gun M-1A1 and the carriage M-1, later it was known as the M-2 and postwar the M-59. The self-propelled version was the M-40.

The 155mm gun has a hydropneumatic variable recoil mechanism and an interrupted screw breech block. It uses the same carriage as the 8in howitzer. When in firing position each of the trails has a spade at each end, the carriage is raised on two jacks. For travelling a two-wheeled limber can be attached to the trails, although in recent years some armies have not used the limber.

The 155mm gun could also be used with the pedestal mount T-6 or firing platform T-6E1 for the coast defence role.

Employment
Austria, Argentina, Greece, Italy, Japan, Jordan, Netherlands, Pakistan, South Korea, Spain, Turkey, Yugoslavia.

155mm Gun in firing position. This shows a German weapon. The weapon is no longer in service with the German Army.

ABOVE
Dutch 155mm Gun being towed by a DAF YA 616 truck

BELOW
155mm guns being used by Japan. Towing vehicle is M-8 tractor.

155mm Auxiliary Propelled Howitzer M-123A1

United States

Calibre: 155mm
Weight: 6350kg
Length Travelling: 7315mm
Width Travelling: 2794mm
Height Travelling: 1942mm
Gun performance etc. is the same as the M-114A1

The M-123A1 (correct designation is Howitzer, Medium, Towed: Auxiliary Propelled, 155mm, M-123A1) consists of the Cannon, 155mm Howitzer: M-1 or M-1A1; Recoil Mechanism: M-6, M-6A2 or M-6B1; Carriage, 155mm Howitzer: M-32.

The weapon is propelled by a hydraulic power unit driven by a Continental Motors Corporation 4 cylinder, 20 HP petrol engine. The 2 wheel drive units are connected to the power unit by 4 hydraulic lines which drive the carriage wheels. A removable caster assembly supports the trails when travelling. Speed whilst travelling only about 7km.p.h. The weapon can still be towed by a truck.

Employment
United States.

155mm Auxiliary Propelled Howitzer M-123A1

155mm Howitzer M-114 and M-114A1

United States

Full designation: Howitzer, Medium, Towed: 155mm M-114 and M-114A1.
Calibre: 155mm
Weight Travelling: 5800kg
Weight Firing: 5760kg
Length Travelling: 7315mm
Width Travelling: 2438mm
Height Travelling: 1800mm
Track: 2070mm
Barrel Length: 3790mm (muzzle to end of breech)
Elevation: -2° $+63^\circ$
Traverse: 25° right and 24° left
Range: 16,600m (HE round charge 7)
Rate of Fire: 1st $\frac{1}{2}$ minute 2 rounds
1st 4 minutes 8 rounds
1st 10 minutes 16 rounds
Ammunition: HE, Chemical, Illuminating, Smoke and Smoke (BE), Gas, Nuclear
Emplacement Time: 5 minutes
Crew: 11
Towing Vehicle: M-4, M-5 or M-6 Tractor 5 ton, 6×6, Truck
Helicopter: CH-47A or CH-47B

This weapon was developed before World War II and first issued in 1942 when it replaced the 155mm M-1918 (this was the French M-1917 Field Howitzer built in the United States). Over 6000 were built and development was by Rock Island Arsenal.

In 1939 the United States was developing a 4·7in gun and a 155mm Howitzer, the latter became the M-114. The 4·7in gun emerged as the 4·5in gun so that it could fire the British shell (British weapon was the 5·5in carriage and 4·5in gun), both carriages were the same.

The designation M-114 was added after World War II, the M-114 consists of the Cannon, 155mm Howitzer M-1 or M-1A1, Recoil Mechanism M-6 (T-2) series and Carriage 155mm Howitzer M-1A1 or M-1A2. The differences are that the M-1A1 Cannon is built of stronger steel. The carriage M-1A1 has a rack and pinion firing jack and the M-1A2 has a screw type firing jack.

The breechblock is of the stepped thread/interrupted screw type and the recoil mechanism is of the hydropneumatic variable type. The trails are of the split type with a small shield on either side of the barrel, the left hand shield top folds down when in the firing position. When in the firing position it rests on three points —both trails and the firing jack. The latter being in front of the shield when in the travelling position, spades are attached to the rear of the trails for firing. For travelling, the trails are locked together and attached to the prime mover.

It is recognisable by its barrel without a muzzle brake, small shields, firing jack and spring type equilibrators each side of the barrel.

FH 155 (L)

This has been modified by Rheinmetall for the German Army, the modifications include a new barrel and muzzle brake, new breech and breech ring and a new shield. In addition new sights and a FCS have been fitted. This is said to have given the weapon a longer range.

M-123A1

This is an auxiliary propelled version and is described elsewhere.

Employment

Austria, Argentina, Belgium, Brazil, Denmark, Germany, Iran, Israel, Italy, Japan, Jordan, Greece, Laos, Libya, Netherlands, Peru, Phillipines, Portugal, Pakistan, South Korea, South Vietnam, Spain, Taiwan, Thailand, Turkey, Yugoslavia.

TOP
Japanese M-114A1 being towed by M-5 tractor

ABOVE
155mm Howitzer of Spanish Army being towed by a Pegaso 3050 6×6 truck

LEFT
155mm Howitzer modified by Rheinmetall. Note the firing jack in position and the muzzle brake and new shield.

106mm Recoilless Rifle M-40A1 and M-40A1C

United States

Calibre: 106mm
Weight: 113kg (rifle only)
130kg (rifle and spotting rifle)
88kg (M-79 mount)
53kg (M-92 mount)
Length: 3400mm
Barrel Length: 2845mm
Elevation: −17° +65° (on mount M-79)
−17° +50° (on mount M-92)
Traverse: 360°
Range: HEAT round, m/v 503 m/s, range 2745m at $6\frac{1}{2}$° elevation
HEP-T round, m/v 498 m/s, range 6876m at $39\frac{1}{2}$° elevation
Effective range is 700–1000m
Crew: 2

The M-40 is a recoilless, air-cooled, breech loaded weapon and its primary role is anti-tank. Development designation was T-170E1 and it was standardised in 1953. It is fitted with a cal ·50 (12·7mm) spotting rifle (model M-8C) mounted on the top of the barrel. Rate of fire is about 5rpm.

The rifle consists of barrel group, breech group, vent assembly and firing cable group:

Barrel group—rifle tube, chamber, quick-breakdown sleeve, spotting rifle front and rear mounting bracket.

Breechblock group—firing pin housing assembly, breechblock hinge and operating lever dog, cocking cam plate, extractor, sear and breechblock operating lever assembly. The breechblock is of the interrupted thread type.

Firing cable group—spotting rifle and 106mm rifle firing cables, rifle firing operating lever, lanyard and rod assembly.

Vent assembly—consisting of ring assembly, recoil compensating ring and retaining screw.

The M-40 can be mounted on any of the following mounts:

Mount M-79—For ground use and for mounting on $\frac{1}{4}$ ton trucks M-38 and M-38A1C. The M-79 has an integral elevating and traversing assembly of the wheelbarrow-tripod type.

Mount M-92—For the M-274 Mule light vehicle, it can also be dismounted from the vehicle and fired from the tripod M-27. The mount M-92 (T-173) does not have an integral base assembly.

Mount M-149E5—This is when 6 of these weapons are mounted on the M-50 ONTOS tracked vehicle used by the USMC. (No longer in service.)

Employment

Australia, Austria (known as the PAK-66 and fitted on a two-wheeled carriage), Brazil (also built in Brazil), Cambodia, Canada, Denmark, France, Germany, Greece, India, Iran, Israel, Italy, Japan (built as Type 60), Jordan, The Netherlands, New Zealand, Norway, Pakistan, Philippines, Singapore, South Vietnam, South Korea, Spain (built in Spain), Switzerland (PAK-58), Taiwan, Thailand, Turkey and Great Britain.

106mm Recoilless Rifle on mount M-79 mounted on a CJ-5 jeep used by the Spanish Army

ABOVE
M-40A1 106mm Recoilless Rifle mounted on the M-274 mechanical mule as used by the United States Marine Corps.

BELOW
106mm Recoilless Rifle built in Spain. This has a longer barrel. The vehicle is the CJ-6 Viasa.

FOOT
Japanese 106mm Recoilless Rifle on a Mitsubishi J54-A jeep

105mm M-102 Howitzer

Howitzer, Light, Towed: 105mm, M-102

United States

Calibre: 105mm
Weight Travelling and Firing: 1450kg
Length Firing: 6718mm
Length Travelling: 5182mm
Width: 1964mm
Height Travelling: 1594mm
Height Firing: 1308mm
Ground Clearance: 330mm
Barrel Length: 3328mm
Elevation: $-5°$ $+75°$
Traverse: 360°
Range: 12,800m
Rate of Fire: 10rpm for first 3 minutes 3rpm sustained
Ammunition: same as M-101A1
Emplacement Time: 4 minutes
Crew: 8
Towing Vehicle: $\frac{3}{4}$ ton, 4×4, or $2\frac{1}{2}$ ton, 6×6, Truck
Helicopter: Chinook

The M-102 was developed by Army Weapons Command to replace the 105mm M-101 Howitzers in airborne and airmobile divisions. It is some 726kg lighter than the M-101, has a slightly better range and can be traversed through 360°. It entered service in 1966.

The M-102 consists of the Cannon, 105mm Howitzer: M-137, Recoil Mechanism, 105mm Howitzer: M-37 and Carriage, 105mm Howitzer: M-31. It has an aluminium box trail and a hydropneumatic variable length recoil, breechblock is vertical sliding wedge, also has spring equilibrators. The weapon can be quickly traversed through 360° as the carriage pivots around the centre of a circular base by means of a roller (Terra-Tire) located at the rear of the trail assembly. The weapon is staked into position, holes being provided in the firing base for this purpose.

The Airmobile Firing Platform XM-6 was developed for use in South East Asia by the United States Army Weapons Command. It consists of a 6706mm square aluminium structure with a plywood deck and adjustable legs at each corner. At the top each leg is a 1220mm square, perforated metal board designed to sink into water, mud or swamp and reach firm ground. The platform took only 6 months to develop and deploy.

Employment
Cambodia, South Vietnam, United States.

XM-164 Experimental 105mm Howitzer

Howitzer, Light, Towed: 105mm, XM-164

United States

Calibre: 105mm
Weight: 1590kg
Length: 7023mm (tube in battery)
5537mm (tube out of battery)
Width: 1894mm
Height: 1473mm
Ground Clearance: 292mm
Elevation: -5° $+75^\circ$
Traverse: $22\frac{1}{2}^\circ$ left and right
Rate of Fire: 3–5rpm (see below)
Range: 12,500m (standard ammunition)
15,300m (extended range ammunition)
Ammunition: see M-102
Crew: 4
Towing Vehicle: 4×4

In 1964 the United States Marine Corps required a replacement for its M-101 Howitzers. They tested the standard United States Army gun, the M-102. This was however, found to be unsatisfactory.

Two weapons were designed. The first, the XM-153, was similar to the M-101 except that it has a longer barrel and a breech ring, extensive use was made of light alloys. The second weapon was the XM-154, this was the M-101A1 with the longer barrel and stripped down of all non-essential equipment. Neither the XM-153 or XM-154 were satisfactory.

The XM-164 was therefore designed by Rock Island Arsenal. The XM-164 has the same barrel as the M-102 but has a horizontal sliding breechblock, its top carriage split trails, cradle and axle are made of aluminium. It also has a variable recoil mechanism, this permits the weapon to fire 7 rounds in 15 seconds. The muzzle brake has two positions—vertical (passive) and horizontal (active). The XM-164 will go in the USMC LVTP-5 amphibious vehicle, this is made possible because of a quick release lock at the front of the cradle which allows a 1322mm reduction in the overall length of the weapon (by retracting the barrel (tube) from battery).

Employment

Tested late 1966 and early 1967, then trials with the United States Marine Corps March–May, 1967.

Howitzer, auxiliary-propelled 105mm M-101 in travelling position

Howitzer, Auxiliary Propelled 105mm M-101

United States

Calibre: 105mm
Length Overall: 6450mm
Width Overall: 2794mm
Height Overall: 1650mm
Tread: 2094mm
Wheelbase (caster trailing): 4254mm
Turning Radius (inside): 3962mm
Weight: 2948kg
Ground Clearance: 254mm
Maximum Speed: 24km.p.h
Maximum Gradient: 60%
Side Slope: 40%
Range: 40km
Fuel: 22 litres
For performance of the weapon refer to M-101 weapon.

The auxiliary-propelled, 105mm, M-101 Howitzer with major/minor wheels was developed by the Lockheed Aircraft Service Company under contract to the United States Army Weapons Command.

The propulsion system provides local ground mobility for tube artillery air-lifted to remote firing sites. With the major/minor wheel feature, the weapon incorporates a self-recovery capability to prevent immobilisation in adverse terrain.

The engine used is a military standard 4A084 (3600rpm), hydraulic pump used is a Sunstrand Series 20, variable displacement (2 used, 2000rpm), hydraulic motor is a Sunstrand Series 20, fixed displacement (2 used maximum motor speed 2000rpm). The engine and caster is on the right trail and the drivers seat on the left trail.

The weapon, crew and 907kg of ammunition can be lifted by the CH-47 helicopter.

Lockheed has also suggested a similar application to be adopted to the XM-204 Howitzer, in this case the weapon would be on a carrier with two major/minor arrangements, i.e. one at each end of the vehicle.

Employment
Trials only.

105mm Recoilless Rifle M-27 and M-27A1
United States

Calibre: 105mm
Weight: 165kg (rifle only)
153kg (mount and adaptor for M-75)
123kg (mount and adaptor for M-75A1)
Length: 3410mm
Barrel Length: 2718mm
Elevation: −13° +50°
Traverse: 60° (M-75)
80° (M-75A1)
Range and Ammunition: HE m/v 341 m/s range 8560m at 43° elevation
WP m/v 341 m/s range 8450m at 43° elevation
HEAT (M-324) m/v 381 m/s range 3200m at 9° elevation
HEAT (M-341) m/v 503 m/s range 2740m at 6° elevation
HEP m/v 385 m/s range 3200m at 9° elevation
HEP-T m/v 515 m/s range 6970m at 40° elevation
Effective Range: 1000m
Crew: 5, including 2 ammunition members

This is a portable, air-cooled, single-loading, recoilless weapon that fires fixed rounds. Its development designation was T-19. Primary role of the M-27 is anti-tank. Breechblock is of the interrupted thread type.

The rifle consists of a tube, firing cable, fulcrum ring, breechblock operating lever assembly, chamber, trunnion assembly, trunnion ring and trigger block. The mount consists of the equilibrator and elevating mechanism assembly, travelling lock assembly and top and bottom carriage assembly. With the use of an adapter the mount is designed for truck mounting, the adapter being attached to the bottom carriage. It can be mounted on a variety of vehicles including the M-38 or M-151 Jeeps or the M-29 Cargo Carrier.

The difference between the mount M-75 and M-75A1 is in the adapter. The mount M-75A1 has an aluminium adapter light enough to be handled by one man and in addition can be attached to three types of jeep. The mount M-75 has a steel adapter. A forcing cone was added in the chamber of the rifle M-27 making it the M-27A1. Telescope fitted is the M-94 or M-90C. It can also be mounted on the Carriage, Rifle, 105mm M-22 (T-47), this can be towed by a 4×4 truck and has large tyres.

Employment
France, Israel, Japan, Morocco, Yugoslavia.

105mm Recoilless Rifle M-27

M-101A1 105mm Howitzer United States

Calibre: 105mm
Weight Firing: 2222kg
Weight Travelling: 2258kg
Length Travelling: 5994mm
Width Travelling: 2159mm
Width Firing: 3657mm
Height Travelling: 1974mm
Track: 1778mm
Barrel Length: 2592mm
Elevation: -5° $+66^\circ$
Traverse: 23° left and right
Range: 11,000m
Ammunition: HE (M-1), HEAT-T (M-67), Blank (M-395), Empty (M-1), Dummy (M-14), HEP-T (M-327), Smoke (M-84A1 and M-84B1), WP (M-60), Chemical (M-360), Illuminating (M-314), TP-T (M-67), TP (M-71), HE (M-3), HE (M-413), HE (M-444), Leaflet (M-488), Gas, CS (XM-629), Beehive (XM-546), also RAP (XM-482, XM-548), HEAT (XM-622)
Armour Penetration: 102mm at 1500m (HEAT round)
Rate of Fire: 1 ½ minute—8rpm
1st 4 minutes—4rpm
1st 10 minutes—3rpm
or 100 rounds per hour
Emplacement Time: 3 minutes
Crew: 8
Towing Vehicle: M-5 Tractor
M-35 or M-135, 6×6, Truck
Airportable: CH-47A/CH-47B Helicopter

The M-101A1 consists of Cannon, 105mm Howitzer, M-2A1, Recoil Mechanism M-2 and Carriage M-2A2. The basic weapon was developed well before the start of World War II, first being known as the gun and carriage M-1 (1928), the weapon entered production in 1939 but by 30th June 1940 only 14 had been built, production was completed in 1953 by which time 10202 had been built.

The weapon has a horizontal sliding wedge breechblock and a hydropneumatic recoil mechanism (constant recoil). Life of the barrel is 20,000 rounds. There was also an experimental auxiliary propelled version called the XM-124.

The 105mm gun was mounted on many self-propelled chassis including the M-7 (Priest), M-37 and T-19 (Half Track), the 105mm cannon was also developed as the T-7 aircraft gun. Also developed during the war was the M-3 105mm Howitzer. This was used for jungle warfare and was basically a 105mm barrel shortened by 686mm and mounted on the 75mm Field Howitzer Carriage, range was 6400m and total weight 1134kg. The carriage was the M-3A1. None of these are known to be in service.

The German firm of Rheinmetall have modified a number of 105mm Howitzers for the German Army, these modifications include a new barrel and new breech ring, new sights and a muzzle brake, range is said to be increased to 14,100m and weight of the weapon is now 2500kg.

The American Lockheed Aircraft Service Company has developed an auxiliary self-propelled version of the 105mm Howitzer, and there is a separate entry for this version.

Employment
Australia, Austria, Argentina, Belgium, Brazil, Bangla Desh, Bolivia, Chile, Cambodia, Colombia, Denmark, Dominican Republic, France, Greece, Guatemala, Haiti, Italy, Jordan, Indonesia, Iran, Japan, Laos, Libya, Liberia, Morocco, Mexico, Netherlands, Norway, Pakistan, Peru, Portugal, Philippines, Pathet Lao, Spain, South Korea, Taiwan, Turkey, Yugoslavia, Thailand, Israel, Vietnam (North and South), Germany, Venezuela

3in Anti-Tank Gun M-5 (76mm)

Entered production in 1942 and production completed 1944. This was T-9 3in A/A Gun fitted with a modified 105mm Howitzer breech ring, recoil mechanism and carriage, became 3in Anti-Tank Gun T-10, and was standardised as the M-5 in 1941. Traverse 22½° left and right, maximum range 14,700m, weight 2170kg, rate of fire 20rpm, time to emplace 3 minutes, towed by a M-3 Half Track or similar vehicle. Reported to be used by Yugoslavia.

TOP
Norwegian 105mm Howitzers being prepared for firing

ABOVE
The modified German 105mm Howitzer. Note the barrel.

BELOW
This is the American 105mm Howitzer modified by the French DTAT. The modifications include replacing the barrel by a 30 calibre barrel with a US powder chamber equipped with a muzzle brake, and installing a pneumatic equilibrator. Maximum range with 105mm Mk 63 shell is 15,000m or 11,500m with 105mm shell HEM1. At the time of writing it is for trials only.

75mm Pack Howitzer M-116 United States

Calibre: 75mm
Weight Travelling/Firing: 653kg
Length Travelling: 3658mm
Width Travelling: 1194mm
Height Travelling: 940m
Barrel Length: 1500mm (including breech ring)
Elevation: −5° +45°
Traverse: 3° (left and right)
Maximum Range: 8500m (HE round)
Rate of Fire: 22rpm maximum
3–5rpm sustained
Ammunition: HE charge 1 m/v 213 m/s
HE charge 4 m/v 381 m/s
Also HEAT, Smoke and Chemical rounds
Crew: 4
Towing Vehicle: M-37, 4×4, truck or $\frac{1}{4}$ ton truck

Development of this weapon dates from before World War II. The first carriage was the M-1, this had box trails and wooden spoked wheels and could be broken down into 6 loads. This was followed by a carriage with pneumatic tyres, small shield and split trails. The later model was the M-8 carriage, this had a box trail and pneumatic tyres. It can be broken down into 8 components for mule transport and it takes about 3 minutes to bring into action from this broken stage. In addition it could also be delivered by parachute. In the latter case it is broken down into 9 components and packed into a para-crate, this takes some 7 minutes to be uncrated, reassembled and emplaced.

After the war the M-8 became the M-116, this consists of: Cannon, 75mm Pack Howitzer M-1A1, Recoil Mechanism M-1A6, M-1A7 or M-1A8 on Carriage, Howitzer (Pack) M-8. There is also a saluting version of this weapon the M-120, which can only fire blank rounds, welding rod has been used to prevent rounds being placed in the chamber.

The M-116 has a horizontal sliding wedge breechblock and a hydropneumatic recoil mechanism.

Employment

China, Cuba, Brazil, Haiti, India, Iran, Japan, Honduras, Laos, North Vietnam, Oman, Pakistan, South Vietnam, Taiwan, Thailand.

75mm Recoilless Rifle M-20 United States

Calibre: 75mm
Weight: 52kg (rifle only)
67kg (rifle and mount M-74)
Length: 2083mm (rifle)
Ammunition: projectile weight is 6·53kg
HEAT-T m/v 305 m/s, range 3200m at 12° elevation
HEP-T m/v 427 m/s, range 6560m at 42° elevation
HE (TP and WP) m/v 302 m/s, range 6360m at 43° elevation
Crew: 2–3

Design of this weapon started in March 1944 and the first pilot was completed and test fired in September 1944, development designation was T-25. The first production order for these went to the Miller Printing Machinery Company of Pittsburgh, and production was under way by March 1945.

It has a rifled barrel and is of the air-cooled type and fires fixed rounds, it is capable of direct or indirect fire, sight fitted is the M-17A1.

The rifle consists of a barrel group and a breech mechanism group. The breech is of the interrupted thread type. The barrel group consists of a tube, tube handle, mounting bracket, hinge block and vent bushing. The breech mechanism group is attached to the rear end of the barrel group and controls the opening and closing of the breech and the firing of the ammunition. The breech mechanism group consists of the breech operating handle housing, the trigger and firing components of the rifle, the closed breech-block, breechblock hinge and the extractor assembly.

The M-20 can be mounted on either the M-1917A2 or M-74 Mounts, both being of the tripod type. The M-1917A2 has a weight of 24kg, elevation −28° to +65° and a traverse of 2° 49 min, or 360° when free. The M-74 has a weight of 15kg and allows the weapon to traverse through 360°. It is built in China where it is known as the 75mm RR Type 52.

Employment
Austria, Bangla Desh, China, France, Greece, Italy, Japan, Norway, North Yemen, Pakistan, Philippines, Saudi Arabia, South Vietnam, South Korea, Taiwan, Thailand, Turkey, Zaire, Yugoslavia.

BELOW
75mm Pack Howitzer in firing position

FOOT
The 75mm M-20 Recoilless Rifle on the mount M-1917A1

The Haflinger 4×4 vehicle on the right is fitted with a M-18 57mm Recoilless Rifle on Mount M-1917A1

57mm Recoilless Rifle M-18 and M-18A1

United States

Calibre: 57mm
Weight: 18·20kg (rifle only)
20·88kg (including integral mount)
Length: 1562mm
Barrel Length: 1220mm
Ammunition: HE, HEAT (WP and TP), Canister
Range: HEAT round m/v 366 m/s, range 1800m at $6\frac{1}{2}°$ elevation
WP (HEAT) round, range 4000m at 39° elevation
HE round, range 4500m at 40° elevation
Crew: 1–2

Development of American Recoilless Rifles started early in 1943 and the first pilot weapon was completed in July 1943, the first actual weapon was the 57mm T-15E1 which was completed in October 1943 and tested in November 1943. The first production order went to the Dominion Engineering Works in Canada. It is built in China (Communist) as the 57mm RR Type 36.

It can be fired from the shoulder or a ground mount. The rifle is composed of six parts—barrel group, breech mechanism group (breechblock is of the interrupted lug type), firing cable group, trigger mechanism group, bipod assembly and an extendable handle assembly on the right hand side of the barrel.

The M-18 and M-18A1 differ in that the breechblock cover on the M-18 has been replaced on the M-18A1 by the breechblock operating lever from which the breechblock handle projects. In addition the M-18A1 has an improved sear and striker assembly.

It is recognisable as it has a handle on the RHS of the barrel, this projects horizontally to the right of the barrel, there is also a support rod, this is under the barrel and drops vertically downwards.

The M-18 is for direct fire only and is provided with a telescope sight. It can also be fitted on the Mount, Tripod, Rifle, M-1917A2. This has a weight of 34kg, traverse of 360° and elevation of −28° to +65°. In addition it can be fitted to the Mount M-74 which weighs 13kg and has a traverse of 360°.

A very similar model was the T-15E16, this was limited production only and was similar to the M-18A1 except that it had a different firing linkage group and a different arrangement of the safety lever assembly on the RHS.

Employment
Austria, China, Cuba, Cyprus, France, Greece, Honduras, India, Italy, Japan, Laos, Norway, South Korea, South Vietnam, Taiwan, Turkey, Thailand, Zaire, Yugoslavia.

120mm Anti-Aircraft Gun M-1 United States

Calibre: 120mm
Weight Firing: 24,289kg
Weight Travelling: 27,897kg
Length Travelling: 9372mm
Width Travelling: 3136mm
Height Travelling: 3149mm
Wheelbase: 4724mm
Ground Clearance: 381mm
Barrel Length: 7391mm
Elevation: -5° $+80^\circ$
Traverse: 360°
Range: 24,780m Horizontal
14,440m Vertical (effective)
Rate of Fire: 10–15rpm
Ammunition: Projectile weight 22·68kg
HE (M-73), AP-T (M-358), HVAP-T (T-144), Smoke (T-16E2), Canister
Muzzle Velocity 944/1066 m/s
Time to Emplace: 40–90 minutes
Towing Vehicle: M-6 Tractor

This weapon was originally the 4·7in M-1 (T-1) and was built during World War II, some 550 were built by the end of the war. The gun is mounted on mount M-1 which has two bogies, each with four dual wheels (in pairs). These are removed before firing and it rests on four outriggers, each having a hydraulic jack.

It has an automatic rammer and a fuse setter and is power driven in elevation and azimuth. The breechblock is vertical sliding, the recoil mechanism is hydropneumatic and variable. Postwar the weapon was deployed in batteries of 4 guns and controlled by the M-33D Fire Control System. Later the gun was designated the M-1A3 and the mount M-1A2.

Employment
No longer used by the United States. Unconfirmed reports indicate that it may be used by Israel.

120mm Anti-Aircraft Gun M-1

90mm M-117 Anti-Aircraft Guns in travelling order. Photograph was taken in Spain.

90mm Anti-Aircraft Gun M-117 United States

Calibre: 90mm
Weight Travelling: 8626kg
Length Travelling: 6350mm
Width Travelling: 2565mm
Height Travelling: 2845mm
Track: 2222mm
Barrel Length: 4724mm
(muzzle to rear face of breech ring)
Elevation: $-5°$ $+80°$
Traverse: 360°
Rate of Fire: 22rpm
Range: 17,000m (vertical)
10,000m (horizontal)
Ammunition: Fixed rounds, HE m/v 823 m/s range 11,790m limited by a 30 second MT fuse. 17,000m with PD fuse. APC-T m/v 823 m/s range 19,000m. HAVP-T range 14,000m. Smoke range 17,800m.
Towing Vehicle: M-4 Tractor or 6×6 $7\frac{1}{2}$ ton Truck

The M-117 was previously known as the M-1, the M-117 designation coming after the last war. The gun and mount were standardised in 1940 and it replaced the 3in M-1918 anti-aircraft gun. The M-117 consists of the 90mm gun M-1A2 or M-1A3 (this can be retubed), Mount M-1A2 and M-1 series Recoil Mechanism. It can be used in both the anti-aircraft and ground roles.

The breech mechanism is of the vertical sliding type and is semi-automatic, and the recoil mechanism is of the hydro-pneumatic type with a variable recoil mechanism. It has both manual and powered elevating and traversing mechanisms.

The mount consists of a carriage, with a single four-wheel axle, and outriggers, which fold vertically. Platforms for crew members are carried on the towbar in travelling order. The weapon takes 7 minutes to bring into action or 20 minutes if the fire control equipment is to be set up. The Fuse-Setter Rammer M-20 is an integral part of the mount. The weapon can also be mounted on a M-3 Pedestal Mount when used in a fixed installation.

Employment
Greece, Spain, Turkey.

M-118 in travelling position

90mm Anti-Aircraft Gun M-118 United States

Calibre: 90mm
Weight: 14,650kg
Length Travelling: 8990mm
Width Travelling: 2620mm
Height Travelling: 3073mm
Barrel Length: 4496mm
Elevation: −10° +80°
Traverse: 360°
Rate of Fire: 23rpm—sustained
28rpm—rapid
Ammunition: fixed rounds: APC-T, Smoke WP, HE, HVAP, T-P.
Range: APC-T m/v 854 m/s, 19,000m
WP m/v 824 m/s, 17,900m
HE m/v 824 m/s, 17,900m
Towing Vehicle: Truck or M-4 Tractor

The M-118 was previously known as the M-2 and was developed from the earlier M-117 during World War II. It was called the Triple Threat Gun as it could be used for anti-aircraft, anti-tank or as field artillery.

The M-118 consists of the 90mm Cannon M-2A1 or M-2A2, Recoil Mechanism M-17 series, Mount M-2A1 and the fuse-setter rammer M-20. The M-20 sets the projectile fuses according to the fire control system or director data and rams rounds into the gun chamber at high speed.

It has a vertical sliding wedge breech-block and a hydropneumatic recoil mechanism. The breech is opened automatically in counter-recoil after firing the first round and closed automatically when the round is rammed. A telescope is provided for use against ground targets.

The weapon is mounted on a four-wheeled carriage. When in the firing position the bogies are removed and the outriggers used. The front and rear outriggers fold upwards and the side ones go along either side of the carriage. The weapon can however be fired when the bogies and wheels are attached.

Employment
Brazil, Greece, Japan, Pakistan, Turkey.

75mm Anti-Aircraft Gun M-51 in travelling order. Note radar scanner on the left side of the barrel is folded down.

75mm Anti-Aircraft Gun M-51 United States

Calibre: 75mm
Weight Travelling: 9480kg
Weight Firing: 8750kg
Barrel Length: 3991mm (muzzle to face of breech)
Elevation: -6° $+85^\circ$
Traverse: 360°
Range: 13,000m (horizontal)
9,000m (vertical)
Ammunition: HE (proximity fuse), m/v 854 m/s, shell wt 5·7kg.
Rate of Fire: 45rpm
Towing Vehicle: M-8A1 Tractor

The 75mm Anti-Aircraft Gun M-51 was developed from the end of World War II and is also known as the Skysweeper. It consists of 75mm Cannon M-35, Recoil Mechanism M-39, Mount M-84 and the M-38 Fire Control System.

It can be used both in the anti-aircraft and ground roles. For the anti-aircraft role it is provided with a tracking radar mounted on the left hand side of the weapon. This has a searching range of 22,000m. Optical sights are fitted for use against ground targets.

The weapon is mounted on a four-wheeled carriage, the axles and wheels being removed when in the firing position. It is then supported on four retractable outriggers, two of which are vertical when travelling and the other two horizontal either side of the carriage.

The recoil system is hydropneumatic and the recoil length is variable. The breech mechanism is automatic and has a vertical sliding breechblock. The high rate of fire being achieved by having two revolver type magazines at the rear of the weapon.

Employment
Greece, Israel, Japan, Turkey.

40mm Light Anti-Aircraft Gun M-1
United States

Calibre: 40mm
Weight Travelling: 2520kg
Length Travelling: 5727mm
Width Travelling: 2020mm
Height Travelling: 1830mm
Wheelbase: 3200mm
Ground Clearance: 370mm
Track: 1406mm
Barrel Length: 3266mm
Elevation: −6° +90°
Traverse: 360°
Range: 4760m maximum horizontal
4600m maximum vertical
2560m effective vertical
Ammunition: HE-T, AP-T
HE wt ·84kg, m/v 875 m/s
Rate of Fire: 140rpm (cyclic)
70rpm (practical)
Armour Penetration: 50mm at 900m (30°)
Crew: 4–6 men
Towing Vehicle: 4×4 or 6×6 Truck

In 1940 the United States Army obtained a 40mm Bofors gun from Great Britain and the United States Navy obtained one from Sweden. The weapon was subsequently built in the United States, the Army having the air-cooled mobile gun and the Navy the water-cooled gun in a twin mount.

The first pilot US Army guns were completed in July 1941 and the weapon was standardised as the 40mm gun M-1 in April 1941. The first production contract was signed in July 1941. The American-built guns differed in a number of ways from the Swedish-built version, for example the chassis was welded instead of riveted, and in numerous other ways the Americans reduced the manufacturing time of this weapon.

The M-1 carriage was for Lend Lease, M-2 for the United States Forces, M-2A1 had new gears with higher gear ratios, the M-5 was a special version for jungle warfare, this was lighter, had only two wheels, was narrower and could be carried by the transport aircraft then in service.

40mm Anti-Aircraft Gun

Various self-propelled models were built including the M-19 and the M-42 (postwar and still in service in many countries) in addition it was mounted on half tracks and many other experimental mounts.

Employment
Brazil, France, Greece, Iran, Israel, Japan, Norway, Pakistan, South Vietnam, Taiwan, Thailand, Turkey.

M-167, Vulcan Air Defence System, Towed

United States

Calibre: 20mm
Weight Travelling: 1588kg
Weight Firing: 1588kg
Length Travelling: 4700mm
Length Firing: 4040mm
Width Travelling: 1980mm
Width Firing: 2360mm
Height Travelling: 2032mm
Height Firing: 1764mm
Track: 1764mm
Barrel Length: 1524mm
Elevation: −5° +80°
Traverse: 360°
Radar Range: 5,000m
Rate of Fire: 3000rpm (high)
1000rpm (low)
Ammunition Supply: 500 ready rounds
Ammunition Types: M-51 Dummy
M-55A2 Target Practice
M-52 Armour Piercing Incendiary Tracer
M-53 Armour Piercing Incendiary
M-54 High Pressure Test
M-56A3 High Explosive Incendiary
M-220 Target Practice Tracer
XM-246 HE Incendiary Tracer
Self Destruct (under development)
Crew: 1 (on weapon)
Towing Vehicle: M-715, 1¼ ton, 4×4, Truck
M-37, ¾ ton, 4×4, Truck

The Vulcan Air Defence System was selected by the United States Army after tests in 1964 and 1965, to arm the new Composite Air Defence Battalions. The Vulcan system is in service in two roles: the M-167 towed or the M-163 on the M-113A1 chassis (self-propelled). The Vulcan system was developed by the Armament Systems Department of the General Electric Company of Burlington, Vermont.

The system is made up of the 20mm six-barrelled M-61A1 Vulcan gun, a linked ammunition feed system (on the left hand side of the weapon), and a fire control system. All mounted in an electrically-powered turret, slewing rates are—azimuth 60° a second and elevation 45° a second. The towed system is provided with an auxiliary power generator mounted on the chassis.

The fire control system consists of a gyro lead-computing gun-sight, a range only radar (mounted on the right hand side), and a sight current generator. The gunner visually acquires and tracks the target. The radar supplies range and range rate data to the sight current generator. These inputs are converted to proper current for use in the sight. With this current the sight computes the correct gun lead angle and adds the required super elevation. The system uses the XM-61 gyro lead-computing gun sight.

The high rate of fire is used against subsonic aircraft and the lower rate of fire for ground targets such as personnel and light AFVs. The gunner can select 10, 30, 60 or 100 round bursts.

The Vulcan gun is also being developed for the United States Navy under the designation CIWS (Close-In Weapons System) for use against ship launched missiles.

Employment
United States Army.

TOP RIGHT
Detailed drawing of the Vulcan Towed Gun M-167. Note the outriggers in position.

RIGHT
M-167 in firing position

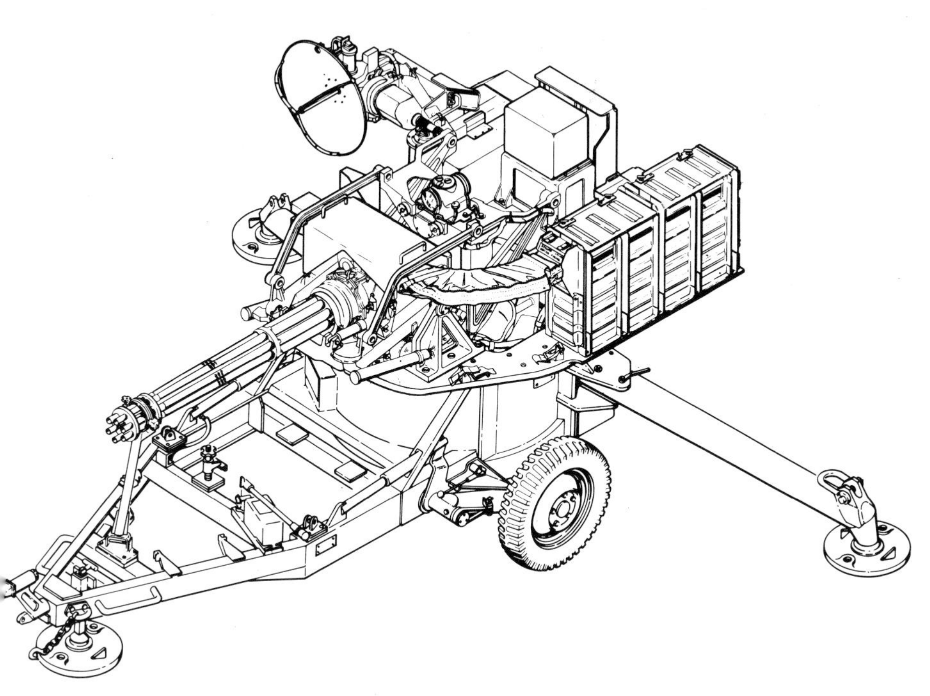

Mount Gun: Trailer, Multiple Cal 50 MG, M-55

United States

Calibre: 12·7mm
Weight: 975kg (M-45C Mount)
363kg (M-20 trailer)
Length: 2890mm
Width: 2090mm
Height Travelling: 1606mm
Height Without Wheels: 1428mm
Ground Clearance: 178mm
Gradient: 60%
Fording: 460mm
Elevation: $-10°$ $+90°$
Traverse: 360°
Rate of Fire: 450/555rpm per barrel
Range: API 5800m, but effective range is much smaller about 1000m
Ammunition: Ball (m/v 892 m/s), AP, API (m/v 929 m/s), API-T, Incendiary, Tracer, Blank and Dummy
Crew: 1–3
Towing Vehicle: $2\frac{1}{2}$ ton, 6×6, Truck

M-55 Quad Cal ·50 mount in travelling order

The M-55 system consists of the Mount M-45C and the Trailer, 1 Ton, Two-Wheel, Machinegun Mount, M-20. The mount M-45C is power driven and semi-armoured. A power charger produces electrical current to be stored in two 6 volt batteries and the electrical system operates from these batteries. The guns have an azimuth speed of 0–60° a second and in elevation 0–60° a second.

The guns used are the M-2 Heavy Machine Gun, 4 of these are used, mounted two each side. Early models had four ammunition chests (2 each side), each of the chests held 200 rounds of ammunition, later models had the chests replaced by ammunition box trays. The M-20 Mount has only two small aircraft type tyres and can only be towed at speeds of up to 16km.p.h. on roads, it is normally carried in a rear of a $2\frac{1}{2}$ton, 6×6, Cargo Truck M-35, this vehicle is fitted with equipment enabling the weapon to be deployed off the vehicle. When on the ground the wheels are removed and the weapon is supported on three jacks, one at the drawbar and two at the rear.

An earlier version of this system was the M-51, this consisted of the M-45 mount and the M-17 trailer, the latter had four vehicle type tyres. During World War II these weapon and Mount M-45 were mounted on the US half track, the system was then known as the M-16 or M-17, depending on the half track used, this vehicle is also still in use. The M-45 was developed from the earlier twin ·50 MG mount, the M-45 was also known as the Maxon.

Employment
Italy, Japan, Jordan, Pakistan, Portugal, Spain, Yugoslavia. The United States and South Vietnam have made extensive use of the M-55 in Vietnam. In this case the M-55 was fitted with a bigger shield and mounted on the rear of a 6×6 truck and used for convoy escort duties.

107mm Mortar M-30 (4·2in) in firing position

107mm Mortar M-30 (4·2in) United States

Calibre: 107mm
Barrel Weight: 71kg
Baseplate (M-24): 100kg
Baseplate (Alternative manufacture): 104kg
Mount M-24A1 (w/o baseplate): 132kg
146kg (alternative manufacture)
Complete Weight: 295kg
Barrel Length: 1730mm
Elevation: 45° to 59°
Traverse: 7° left and right
Rate of Fire: 5rpm normal
20rpm maximum
Ammunition: HE, Illuminating, Smoke (PWP and WP), Chemical (CG, H, HD and HT)
Range:
HE (M-329 or M-329B1) full charge: 5420m m/v 293 m/s elevation 45°
HE (M-3A1, M-3): 4600m m/v 257 m/s elevation 45°
Chemical (M-2 and M-2A1): 4600m m/v 257 m/s elevation 45°
Illuminating (M-335): 4800m m/v 283 m/s elevation 51°
Crew: 5–6

The mortar consists of the Cannon M-30 on Mount M-24 or M-24A1. It was developed by the Chemical Corps, development designation was T-104 and it started to replace the old M-2 Mortar in 1951.

It is a rifled muzzle loaded weapon and breaks down into six parts for transportation: barrel, baseplate, base ring, bridge, rotator and standard.

The barrel is closed at the breech end by a tube cap, this has an integral firing pin, the tube cap is supported and aligned on the baseplate and standard assembly, this is equipped with a screw-type mechanism for elevation and traverse.

The Mount M-24 or M-24A1 consists of a rotator assembly, a bridge assembly, a standard assembly and a baseplate. The M-24A1 baseplate is a one piece steel one and the M-24 one is of magnesium with an inner baseplate and an outer baseplate ring.

The M-30 is also mounted in the M-106 and M-106A1 tracked vehicles, these being modifications of the basic M-113 and M-113A1.

Employment
Austria, United States, Zaire and most of the countries that use the older M-2 Mortar.

107mm Mortar M-2 (4·2in) United States

Calibre: 107mm
Total Weight: 151kg
Barrel Weight: 48kg
Standard Weight: 24kg
Baseplate Weight: 79kg
Elevation: +45° to +60°
Traverse: 7° left and right
Range: 516–4022m

The M-2 Mortar consists of a barrel, standard and baseplate. The barrel is rifled, and the baseplate rectangular and is of ribbed steel construction. The standard looks like an inverted 'T' piece and this is connected to the baseplate by connecting rods. It has a single vertical column that houses the elevating mechanism. The weapon is easily recognisable by its baseplate and the connecting rods connecting the standard and the baseplate.

The M-2 is also reported to have been produced in Spain.

Employment

Belgium, Ethiopia, India, Israel, Italy, Japan, Jordan, Netherlands, Norway, Pakistan, Philippines, South Korea, South Vietnam, Spain, Turkey.

Mortar: 81mm M-29 United States

Calibre: 81mm
Barrel Weight: 12·70kg
Baseplate (inner and outer ring): 21·00kg
Bipod: 18·15kg
Total weight including sight: 52·20kg
Barrel Length: 1295mm
Elevation: +40° to +85°
Traverse: 4° left and right
Ammunition:

Type	*Muzzle Velocity*	*Range*	*Elevation*
HE (M-43A1) and TP (M-43A1)	211 m/s	3350m	45°
HE (M-56 and M-56A1)	174 m/s	2320m	45°
HE (M-362 and M-375A2)	234 m/s	3644m	45°
Illuminating (M-301A1 and M-301A2)	174 m/s	2100m	45°
Smoke FS (M-57 and M-57A1)	174 m/s	2216m	45°
WP	174 m/s	2268m	45°
M-68 Training Round	52 m/s	280m	45°

Rate of Fire: 18rpm (normal)
30rpm (maximum)
Crew 8 (total)

This consists of cannon: 81mm, M-29 and mount mortar: 81mm M-23 or M-23A1. The M-29 was developed from the M-1 and replaced the M-1, the M-29 also uses the same ammunition as the M-1. Development designation of the M-29 was T-106. A more recent model of the M-29 is the M-178 and M-178A1 (M-29E1).

The barrel consists of a tube with a base plug and a fixed firing pin for drop firing and is of the smooth bore type. The exterior of the barrel is helically grooved to reduce weight and this also aids in heat dissipation. The mount consists of a bipod with traversing and elevating mechanisms and a spring type shock absorber above the barrel. There is also a chain between the two legs. The baseplate, which is circular, consists of an inner and outer ring assembly. The M-23 has a cross levelling mechanism consisting of a turn buckle and a clamp arrangement. The baseplate has three baseplate latches. The M-23A1 baseplate has no latches and its cross levelling mechanism is an adjustable sliding one on the left leg.

Employment

Austria, Italy, United States, South Vietnam etc.

81mm Mortar M-1

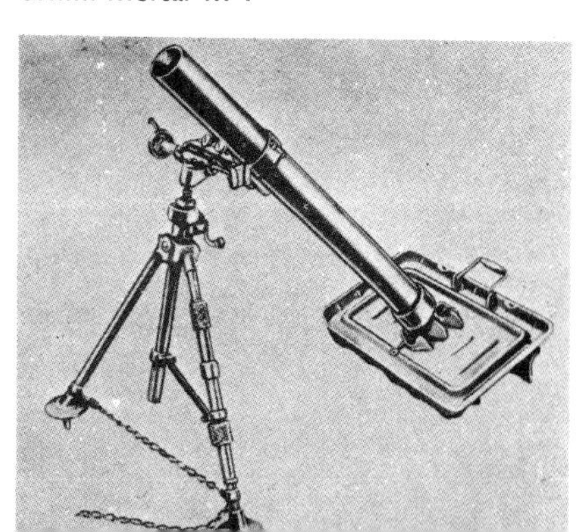

81mm Mortar M-1

United States

Calibre: 81mm
Barrel Weight: 20·2kg
Baseplate Weight: 20·4kg
Total Weight: 62·0kg
Barrel Length: 1265mm
Elevation: +40° to +85°
Traverse: 5° left and right
Rate of Fire: 18rpm (normal), 30rpm (max)
Ammunition: HE, Illuminating, Smoke FS and WP
Range:

Ammunition	*Muzzle Velocity*	*Angle*	*Range*
HE (M-43A1) and TP (M-43A1)	211 m/s	45°	3016m
HE (M-56 and M-56A1)	174 m/s	45°	2317m
HE (M-362)	234 m/s	45°	2467m
Illuminating (M-301A1 and M-301A2)	174 m/s	51°	2102m
Smoke, FS (M-57 and M-57A1)	174 m/s	45°	2216m
Smoke, WP (M-57 and M-57A1)	174 m/s	45°	2568m
Projectile, Training, M-68	53 m/s	45°	283m

Crew: 2–3 men

The 81mm Mortar consists of cannon, 81mm, mortar M-1 and mount, mortar, 81mm, M-4. The M-1 consists of the tube, base cap and firing pin. The base cap is hollowed and threaded to screw onto the barrel. It ends in a spherical projection, flattened on two sides, this fits into and locks in the socket of the base plate. The firing pin is screwed into the basecap against a shoulder, the firing pin protrudes about 1/20th of an inch through the base into the barrel.

The bipod consists of the leg assembly, two chains connect the legs, elevating mechanism assembly and the traversing assembly. The base plate is rectangular and is of pressed steel, to this has been welded a series of ribs and braces, a front flange, three loops, a link, two carrying handles and the socket.

The weapon can be quickly broken down into three units—barrel, baseplate and bipod for easy transportation. The barrel is smooth bore. The mortar is also mounted in the M-4 and M-21 half tracks.

Employment
Australia, Austria, Belgium, Brazil, Cambodia, Cuba, Denmark, Greece, India, Italy, Israel, Indonesia, Japan, Liberia, Luxembourg, Netherlands, Norway, North Yemen, Pakistan, Philippines, South Korea, South Vietnam, Spain, Taiwan, Thailand, Trinidad, Turkey, Yugoslavia, Switzerland.

NOTE: *It has also been reported that the 81mm Mortar M-1 has been built in Belgium, Spain and Yugoslavia.*

81mm Mortar M-29

Rocket Assisted Projectiles (RAP)

United States

RAPs for 105mm and 155mm howitzers were standardised several years ago and are now in service. For example, the M-102 has a maximum range of 12,800m (earlier reports put it at 11,500m which was the same as the M-101), with the RAP it has a range of 15,000m. The projectiles have a solid-propellent rocket motor that provides a 2–2½ second burn. It is also reported that the new projectiles have improved warheads. In the SP role RAPs are being developed for the M-110E2 and are already used with the M-109A1, this has a much longer barrel than the standard M-109.

Terminal Homing

Development of this was started by the United States Navy but the programme has now been taken over by the United States Army. It is said that this could well be a 'breakthrough', according to some American reports.

Post-war American Artillery

During the 1940s America developed many anti-tank guns that did not enter service, for example the 90mm T-133 (smooth bore, built in 1951), 90mm T-8 (1945), 105mm T-8 (1945), as well as guns and howitzers, i.e. the 175mm gun T-45 and the 156mm T-148 Howitzer (1953). In the 1960s was the XM-70, a 115mm Rocket Boosted Artillery System for the US Army and Marine Corps. The M-28 (120mm) and M-29 (155mm) Battle Field Atomic Weapons have been phased out of service, as has the 280mm gun M-65. Other postwar weapons are described in the text. Apart from the M-102, all towed artillery used by the United States is of World War II vintage. The reason is that most American artillery is self-propelled, apart from those used by the airmobile and airborne divisions. Current weapons under development are the XM-204 105mm Howitzer and the XM-198 155mm Howitzer.

105mm Howitzer XM-204

This is being developed to replace the current M-101A1 and M-102 105mm Howitzers at present in service. The first experimental XM-204 was built in 1970, and the engineering phase was approved in 1972, fabrication of the prototypes 1 and 11 began in December, 1972 at Rock Island Arsenal. Few details have been released although the range if reported to be 14,000m, weight about 1814kg and rate of fire is access of 10rpm. The turntable allows the weapon to be quickly traversed through 360°, it is provided with a towing lug under the barrel.

The main feature of the weapon is in its soft recoil system. Development of the soft recoil (or FOOBS—fire out of battery) principle goes back at least to 1962, if not before.

In a normal gun, when it is fired the barrel goes to the rear (recoil) and then forwards to its original position (counter-recoil), this takes about 3 seconds for a 105mm Howitzer. In the XM-204 the barrel is held in position at the rear of the carriage by latches, when the lanyard is pulled these latches are released, this allows the barrel to travel forward. A velocity sensor is mounted on the barrel, this determines when the charge is to be ignited. The barrel then returns to the rear where it is retained again by the latches. This takes 1·4 seconds. The main advantages are the much reduced recoil forces (70 to 75%), this in turn means that no trails are required and the weapon is lighter and shorter.

Employment—Trials

155mm Howitzer XM-198

By 1972, two prototypes had been built of the XM-198 Howitzer and the first had already started its 15,000 round durability test at Camp McCoy, Wisconsin. Eight additional XM-198s are to be built by the US Army Weapons Command. The XM-198 consists of the XM-139 carriage, XM-45 recoil system and XM-199 cannon. Rock Island build the carriage and recoil mechanism, Watervliet the cannon, Frankford Arsenal the fire control equipment and Pitcatinny the ammunition. Rock Island will assemble the weapon before it goes to Camp McCoy.

The XM-198 will, then fully developed, replace the M-114A1 in airmobile and airborne units. It will have a greater range than the M-114A1, rapid shift capability (i.e. traverse) and a higher rate of fire. It will also be lighter although according to one American Army Report the XM-198 weighs 6486kg against the 5760kg of the M-114. The XM-198 will probably enter service in the mid-1970s.

Employment

Trials.

BELOW
XM-204 105mm Howitzer

FOOT
XM-198 155mm Howitzer in firing position

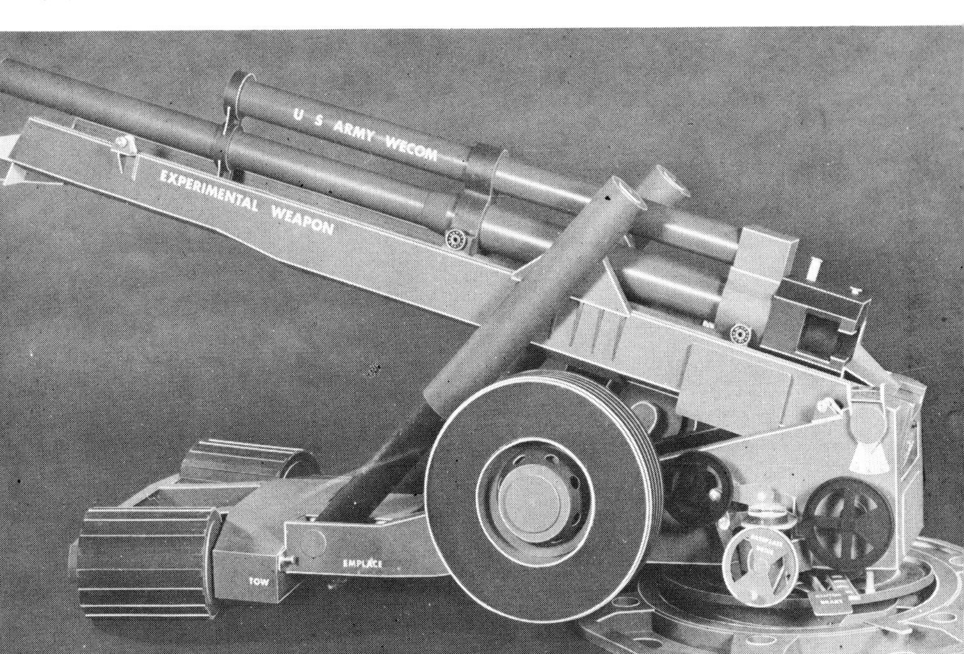

105mm Light Field Howitzer M-1956 (M-56)

Yugoslavia

Calibre: 105mm
Weight: 2180kg
Elevation: $-12° +68°$
Traverse: 52° (total)
Rate of Fire: 6–7rpm (practical)
Range: 13,000m
Barrel Length: 3480mm
Ammunition: HE, HEAT, Smoke, Illuminating. Weight 15kg, m/v 570 m/s
Crew: 6
Towing Vehicle: TAM-4500, 4×4, Truck

This weapon is manufactured in Yugoslavia and is based on the American 105mm M-2 Howitzer.

It is easy to recognise by its multi-baffle muzzle brake and the cylinders that project from the breechblock to midway down the barrel (above and below the barrel). The shield is in two halves, left and right, they are joined by a small rib of metal over the barrel and flaps are provided to protect the axle. The shield is flat at the front and angled at the sides. The carriage has single tyres and split trails with a spade on each trail.

Employment
Yugoslavia.

Rear view of Yugoslav 105mm Light Field Howitzer M-1956

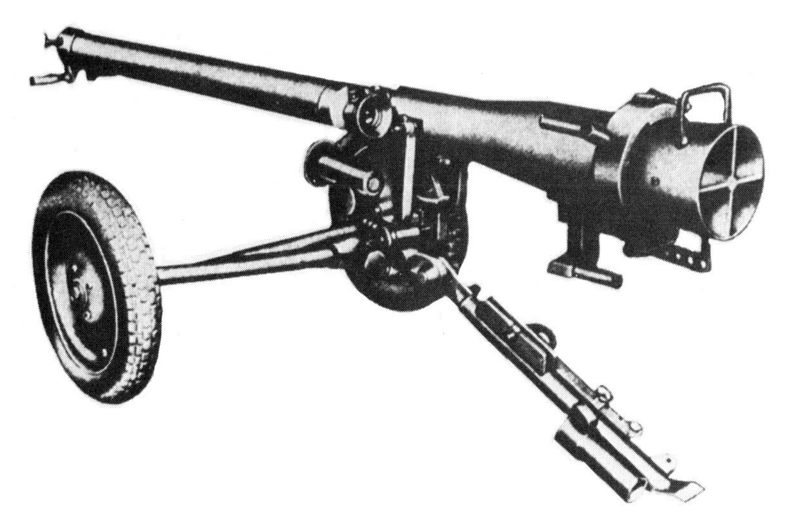

BO-82mm M-60 Recoilless Rifle. Note the stabilising leg and the attachment for fitting the leg to the breech for travelling.

BO-82mm M-60 Recoilless Rifle Yugoslavia

Calibre: 82mm
Weight: 122kg
Length Travelling: 2200mm
Elevation: −20° +35°
Traverse: 360°
Range: 4500m (Indirect Fire)
1000m (Moving Target)
1500m (Stationary Target)
Rate of Fire: 4–5rpm
Ammunition: HEAT weight 7·2kg, m/v 388 m/s
Armour Penetration: 220mm
Crew: 2–5
Towing Vehicle: AR-51 Zastava, 4×4

This weapon is similar in some respects to the Soviet 82mm Recoilless Weapon B-10. The Yugoslav weapon is mounted on a two-wheeled carriage with a stabilizing leg at the rear. The barrel is provided with a handle either side at the front to facilitate handling of the weapon. The sight is on the left hand side of the barrel, there is also a fore sight on the left hand side of the barrel. The M-60 is towed breech first and can also be manhandled into position.

Employment
Yugoslavia.

76mm Howitzer M-1948 (B-1) Yugoslavia

Calibre: 76·2mm
Weight Firing: 705kg
Length Travelling: 2438mm
Width: 1828mm
Height: 1524mm
Barrel Length: 1178mm
Elevation: $-15°$ $+45°$
Traverse: 50° (total)
Range: 9000m
Ammunition: HE, HEAT, Smoke, Weight 6·2kg, m/v 222–420 m/s
Armour Penetration: 102mm at 0°
Rate of Fire: 25rpm (maximum)
6–7rpm (normal)

This weapon can be broken down into four loads for pack transportation. It has folding split trails, the rear half of each trail folds through 180° so that it rests on the forward part of the trail. The shield has wings and splits in two horizontally for increased elevation.

The M-1948 (B-1) is recognisable by its short barrel with a multi-baffle muzzle brake, high shield with a straight top and wings and tyred wheels.

Employment
Yugoslavia, Burma, Ceylon.

105mm Recoilless Rifle M-65 Yugoslavia

No details of the 105mm Recoilless Rifle are available. It does, however, appear to have a spotting rifle on top of the barrel, there is a small shield to the left of the barrel. The suspension system on the carriage is very similar to that used on the 20mm Light Anti-Aircraft gun described below.

20mm Light Anti-Aircraft Gun M-55 (Quad) Yugoslavia

This weapon is the Swiss HSS-603 triple 20mm cannon built under licence in Yugoslavia. The weapon is mounted on a two-wheeled carriage, when in the firing position the wheels are off the ground. The magazines are of the drum type, these are arranged two forward for the left and right barrels and one to the rear of the other two for the centre barrel. Each barrel is provided with a flash hider, these have circular holes in them. Weight of the weapon in firing position is about 1170kg, elevation $-5°$ $+83°$, traverse 360°, the effective range in the anti-aircraft role is 1500m. Crew is six men. Towing vehicle is AR-51, $\frac{1}{4}$ ton, 4×4 truck.

120mm Mortar UBM-52 Yugoslavia

Calibre: 120mm
Weight Travelling: 420kg
Weight Firing: 400kg
Barrel Length: 1290mm
Elevation: $+45°$ to $+85°$
Traverse: 6°
Rate of Fire: 15–25rpm
Shell Weight: Heavy 15·9kg
Light 12·25kg
Ammunition Types: HE, Smoke and Illuminating
Range: 6010m with light shell
4760m with heavy shell
Crew: 5
Towing Vehicle: AR-51 Zastava, 4×4, Truck

This 120mm Mortar is easily recognisable by its medium sized road wheels that have solid tyres, these wheels have 14 triangular holes in each of them. Each wheel is provided with two eyes through which ropes can be put for carrying purposes.

When in travelling position the weapon

ABOVE
Photo shows the Howitzer in firing position and the right upper photograph shows the rear part of the trails folded

LEFT
105mm Yugoslav Recoilless Rifle M-65

BELOW
20mm Light Anti-Aircraft Gun M-55 in firing position

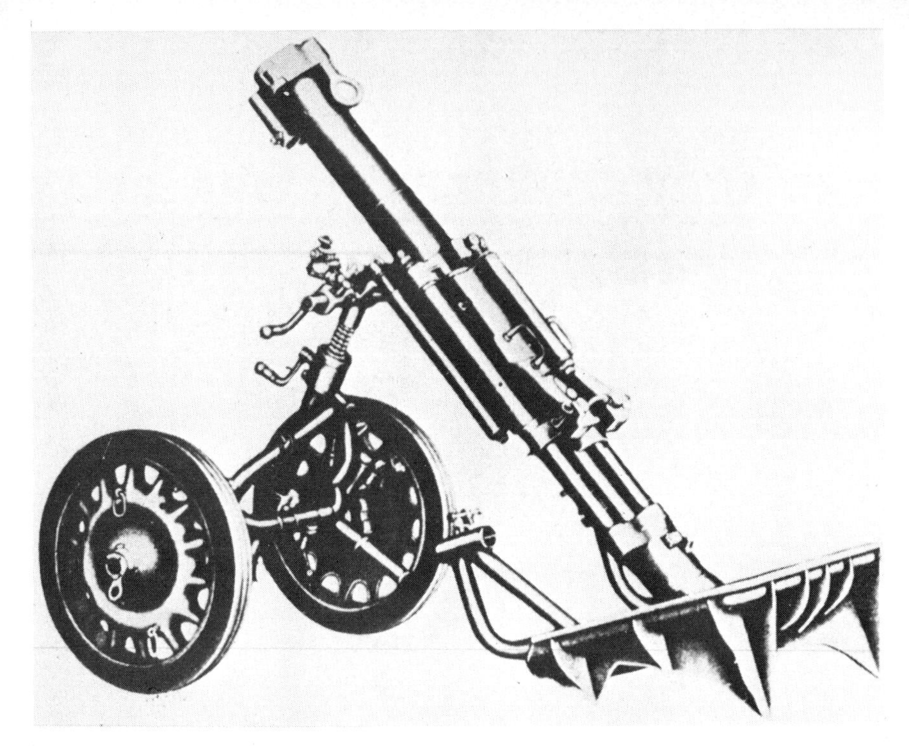

120mm Mortar UBM-52 in firing position

is towed by its muzzle end with the barrel horizontal, the towing eye folds down on top of the barrel when not in use.

When required for firing the link between the webbed baseplate and the axle of the mount is broken enabling the barrel to go to the rear. It is reported that the UBM-52 is the French MO-120 AM 50 120mm Mortar built under licence in Yugoslavia.

Employment
Burma, Yugoslavia.

Other Mortars Used by Yugoslavia

81mm M-31
Reported to be the United States M-1 Mortar built in Yugoslavia.

81mm M-68
Reported to be a French Mortar (Hotchkiss-Brandt model MO-81-61 L) built in Yugoslavia.

There is also reported to be a version of the German World War II Gr.W.34 built in Yugoslavia.

India

It has been reported that India is developing a lightweight mountain pack howitzer. She already produces ammunition and mortars, for example French 81mm and 120mm mortars are built in India.

Japan

In the 1950s Japan built 75mm and 57mm Recoilless Rifles for the United States as well as 81mm mortars. More recently she has built the Swiss 35mm Anti-Aircraft Gun under licence, as well as the 106mm M-40 Recoilless Rifle.

South Africa

Is reported to be building her own artillery, the first of which is said to be a 90mm Field Gun, some reports indicate that this is a modification of the British 25 Pounder, others that it is a weapon of a new design. South Africa already produces much of her own ammunition, up to 160mm rifled and 260mm smooth bore.

International Project (FH-70 and SP-70)

Germany/Great Britain

On 9th August, 1968 an agreement was signed between Great Britain and the Federal German Republic to develop a new 155mm gun. Companies associated with the project are Vickers Limited (Great Britain) and Faun-Werke and Rheinmetall (Germany). It is being developed in two versions: Towed is the FH-70 and Self-Propelled is the SP-70. The towed model will be airportable and probably have an auxiliary engine. No performance figures have been released but it is thought that it will have a range of 24,000m with conventional ammunition and 30,000+ metres with rocket assisted projectiles. Italy has recently joined the programme.

ADDENDA

Soviet Union

Another new Soviet weapon has recently been observed, this is the SPG-9. This has a calibre of 76mm and is similar to American recoilless rifles, i.e. the 75mm M-20. The SPG-9 is mounted on a tripod with the sight on left side. Weight is approx 55kg, crew of three men and a maximum range of 900m. Maximum rate of fire is 6rpm and its fin stabilised round will penetrate 300mm of armour plate.

This photograph shows a Chinese Communist 57mm Recoilless Rifle Type 36, which is similar to the American 57mm M-18

Tracked Artillery Tractors

Vehicle	**AT-P**	**AT-L**	**AT-S**	**ATS-59**
Length overall	4054mm	5120mm	5870mm	6300mm
Width overall	2335mm	2200mm	2570mm	2800mm
Height overall	1707mm	2180mm	2535mm	2500mm
Weight loaded	6·3t	6·3t	12t	13t
Weight empty	—	4·3t	9t	10t
Armament	1 × 7·62mm MG	nil	nil	nil
Track	2000mm	1875mm	1900mm	2250mm
Track width	300mm	300mm	425mm	425mm
Ground clearance	300mm	300mm	400mm	425mm
Ground pressure	0·56kg/cm²	0·60kg/cm²	0·58kg/cm²	0·52kg/cm²
Engine hp	110	135	250	300
Cylinders	6	4	12	12
Type	Petrol	Diesel	Diesel	Diesel
Gradient	30°	30°	28°	35°
Fording	900mm	600mm	1000mm	1500mm
Ditch crossing	1220mm	1524mm	1450mm	2500mm
Vertical obstacle	600mm	600mm	600mm	1100mm
Cruising range km	300	300	380	350
Maximum speed km/ph	60	42	35	40
Towing capacity	5t	6t	16t	14t
Drive sprocket	FRONT	FRONT	REAR	FRONT
Crew	3+6	3+8	7+10	2+14
Number of road wheels	5	5	8	5
Country	USSR	USSR	USSR	USSR

Full designations are:—

Armoured Tracked Artillery Tractor AT-P

Can tow 85mm, 100mm and 122mm weapons.

Light Tracked Artillery Tractor AT-L

Can tow 57mm AAG, 160mm and 240mm Mortars, 152mm Howitzer.

Medium Tracked Artillery Tractor AT-S

Can tow 100mm AAG, 152mm Howitzer.

Medium Tracked Artillery Tractor ATS-59

Can tow 130mm Field Gun.

Heavy Tracked Artillery Tractor AT-T

Can tow 203mm G/H, 130mm Anti-Aircraft Gun.

Light Tracked Artillery Tractor Csepel K-800

This is a Hungarian copy of the Soviet M-2 tractor and can tow the KS-19 100mm Anti-Aircraft Gun or 152mm Howitzer.

Medium Tracked Artillery Tractor Mazur 300

This was known as the ACS and is based on the Soviet AT-S tractor. Can tow 100mm AAG, 152mm Howitzer.

Tractor, Full Tracked, High Speed, M-5 Series

Developed in World War II, includes M-5, M-5A1, M-5A2, M-5A3, M-5A4, can tow guns of 105 and 155mm calibre.

Tractor, Full Tracked, High Speed, M-4 Series

Developed during World War II, includes M-4, M-4A1, M-4A1C, M-4A2 and M-4C. Can tow 8in Howitzer or 155mm Gun.

Tractor, Full Tracked, High Speed, M-8 Series

Includes M-8A1 and M-8A2. Can tow 75mm Anti-Aircraft Gun or 90mm Anti-Aircraft Gun.

Employment

Communist tractors are used by most

AT-T	K-800	300	M-5s	M-4s	M-8s
6990mm	5000mm	5790mm	4851mm	5156mm	6730mm
3170mm	2400mm	2880mm	2540mm	2463mm	3314mm
2845mm	2200mm	2600mm	2641mm	2514mm	3048mm
20t	6·4t	13·6t	13t	14t	25t
15t	4·6t	10·1t	—	—	17t
nil	nil	nil	1×12·7mm MG	1×12·7mm MG	1×12·7mm HG
2640mm	2100mm	2680mm	2108mm	2032mm	2607mm
508mm	300mm	425mm	288mm	420mm	547mm
425mm	300mm	475mm	500mm	508mm	489mm
0·52kg/cm²	0·45kg/cm²	0·65kg/cm²	0·79kg/cm²	0·53kg/cm²	0·58kg/cm²
415	125	300	207	210	863
12	6	12	6	6	6
Diesel	Diesel	Diesel	Petrol	Petrol	Petrol
30°	30°	28°	28°	30°	30°
750mm	600mm	1000mm	1346mm	1040mm	1070mm
2100mm	1500mm	1400mm	1676mm	1524mm	2133mm
1000mm	500mm	600mm	457mm	736mm	762mm
700	300	350	240	290	290
35	35	55	48	53	64
25t	8t	15t	9t	11·3t	17·7t
Front	Rear	Front	Front	Front	Front
4+14	2+14	9	9	11	9
5	5	5	5	5	6
USSR	Hungary	Poland	USA	USA	USA

Warsaw Pact Forces as well as other Armies, for example: Finland AT-S, Albania AT-T.

American tractors are no longer used by the United States Army but are still found overseas, i.e.: Spain M-4, Japan M-5 and M-8, Yugoslavia M-4,

Soviet AT-L (Modified) Light Artillery Tractor, with five road wheels and no track support rollers. The original AT-L has six smaller road wheels and three track support rollers.

ABOVE
This is the latest Soviet tractor, the M-1970. The photograph shows it with 122mm weapons. Provisional details are length 6350mm, width 2800mm, height 2250mm, weight approx 10tons and a speed of 50km/hr. It is armed with a turret mounted machine gun and is based on PT-76 components.

LEFT
Medium Tracked Artillery Tractor AT-S, of the Finnish Army

BELOW LEFT
Soviet ATS-59 Medium Tracked Artillery Tractor

TOP RIGHT
The Polish Mazur-300 Medium Tracked Artillery Tractor

CENTRE RIGHT
The Soviet AT-T Heavy Tracked Artillery Tractor

RIGHT
M-5 Tractor of the Japanese Self Defence Force towing a 105mm Howitzer. Later models of the M-5 were the M-5A1, M-5A2 and M-5A3.

ABOVE
M-4 tractor of the Spanish Army towing a 8in (203mm) Howitzer

BELOW
M-8E2 Tractor. Note the dozer blade at the front and the tail hoist.

RAP-14 Multiple Rocket System France

The RAP-14 is a new weapons system designed by the French Company CNIM (Constructions Navales et Industrielles de la Méditerranée). It can be towed behind a normal truck, when required for firing four jacks are lowered and the wheels lifted off the ground. It takes only two minutes to take into or out of action as the weapon is fitted with hydraulic equipment.

Total capacity of the launcher is 22 rockets, these being in layers of 6, 5, 6 and 5. Other data is:

Combat Weight: 4500kg
Elevation: 0° +52°
Traverse: 360°
Total time to fire salvo of 22 rounds: 10 seconds
Time to re-load: 1 minute

The above launcher fires the RAP-14 rocket which has a maximum speed in access of Mach 2, the rockets are stabilised by their cross shaped fins. Basic data of the rocket is:

Length: 2000m
Calibre: 140mm
Span: 360mm
Total Rocket Weight: 52kg
Warhead Weight: 19kg
Maximum Range: 16km
Error: 1% of range

Also under development is the RAP-14S rocket which has a range of 20km. The RAP-14 has also been developed for naval use. In this case the mount is the M 18-90 which can launch a total of 18 rockets and is provided with a semi-automatic loading system.

Employment
Trials.

ABOVE
RAP-14 system in firing position

BELOW
The RAP-14 rocket

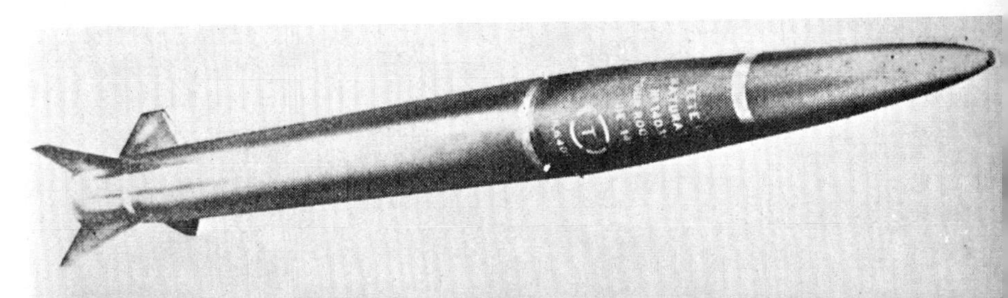

Light Artillery Rocket System (LARS)

Germany

This has been developed in Germany and basically consists of a 6×6, 7ton, Klöckner-Humboldt-Deutz Jupiter truck on which has been mounted 36 launching tubes in two assemblies of 18 each. The cab of the vehicle is armoured, armoured visors protect the cab when the rockets are fired, a 7·62mm machine gun is mounted on the roof of the cab. The system was accepted for service in 1969 and delivery commenced in 1970. Delivery has now been completed.

It is employed as an area saturation weapon with a range of 6000m to 14,000m. The rockets can be fired either as single rounds, part ripple or a complete ripple. The time between the rockets firing is ·5 seconds, the complete 36 rockets can be fired in 18 seconds.

The launcher consists of a lower carriage, an upper carriage and the launcher frames themselves. The launcher has a traverse of 105° each side and can be elevated from 0° to +55°. The length of the launcher tubes is about 3900mm, the Panoramic Sight Type 39 is positioned above and between the two launcher frames. The rockets are launched from within the cab and the vehicle is fitted with automatic rocket testing equipment (REPAG). The crew of three: launcher/commander, aimer and driver all remain in the cab whilst the rockets are being launched. Two jacks are let down at the rear of the vehicle before the rockets are fired.

The artillery rockets have four folding fins and are spin stabilised, they have a calibre of 110mm (including the driving bands), and have a length of 2263mm and weigh 35·15kg. The warhead and fuse weigh 17·20kg. The rocket has a one stage solid fuel motor and burning time is about 2·2 seconds.

According to the manufacturers, the rocket can be fitted with four types of warhead—anti-personnel, smoke, training and anti-tank. Other sources state that the LARS system has the codename of *Dragon Seeder,* and that the anti-personnel rocket is called *Dragoon Seed*. This is said to have some 5000 steel balls and is fitted with a proximity fuse. The anti-tank rocket is believed to be called *Pandora* and this has eight mines for use against tanks, these are fitted with pressure fuses. Another rocket, probably the smoke one, is called *Medusa.*

The system has a complete combat weight of 15,000kg, unloaded weight of 13,500kg, length of vehicle is 7850mm, width 2500mm and height is 2900mm loaded and 3195mm loaded and with the launcher elevated at 15°.

Employment

Only with the German Army. LARS is deployed in batteries of 8 launchers attached to each division.

Front view of LARS with the launcher traversed to the right

BELOW
Rear view of the LARS System. Note the jacks in position.

FOOT
LARS System firing a single rocket

Artillery Rockets
Brazil

Brazil has developed at least two types of artillery rocket:

108-R
This has a total of 16 tubes arranged in two rows of five and the centre one of six. These are mounted on the rear of a 4×4 truck. Range is reported to be 12,000m, the rocket has a HE warhead.

114mm Rocket
Five of these are mounted on a trailer and towed by a truck, range of the rocket is about 25,000m.

Rafale Rocket System
France

This is being developed by the Sociéte Européene de Propulsion. Only brief details have been released and these include a calibre of 142mm, or 300mm with its four folding fins unfolded, weight of 65kg, maximum range of 30,000m with a maximum velocity of 1000 m/s. Warhead would be HE or anti-personnel. It has a solid propellent rocket motor. Probable mount would be one of 30 tubes, and this would take some 20 seconds to fire. The system could be mounted on a vehicle or trailer.

Great Britain

In 1971 the British Army announced that it was developing a Field Artillery Rocket System. Various names have been mentioned, these including *Mark Foil*, and *Foil*, the latter is the more recent one and this also has the designation RS 80. Calibre is reported to be 250mm and range 30/40km, types of warhead would include anti-tank and anti-personnel. It may be being developed in association with Germany and Italy. Britain is also developing a short range anti-personnel mine-laying system called *Ranger*, this consists of 72 tube type magazines each containing 20 minlets.

Italy

For some years Italy has used 100mm rockets, these are mounted on the rear of 4×4 light trucks in multiples of 24. The following are some of the more recent artillery rockets:

Attala Rocket:
Calibre 80mm and a maximum range of 800m.

Super-Attala
Calibre 104mm and a maximum range of 2000m.

Samurai:
Calibre of 76mm and a maximum range of 18,000m.

In addition there are at least two other artillery rockets under development including the **Brenda BR.51.GS.** This system consists of rockets with a calibre of 158mm and a range of 24,000m, weight of 123kg and four types of warhead are being developed. It will be mounted on a tracked chassis in multiple tubes. Italy is also involved in the British RS 80.

Rear view of the Type 30 Rocket on its launcher vehicle

Japan

Japan has developed a number of rockets since World War II. One of these is the **Type 30** rocket which has been developed by the Space and Aeronautical Department of the Nissan Motor Company Limited. The Type 30 has a calibre of 300mm, a length of 4500 mm and a range of some 30,000m. No other details have been released. Two of these missiles are mounted on a Japanese Hino, 6×6, Truck, this is provided with jacks to stabilise the vehicle when the rockets are fired. Another Hino truck carries six reserve rockets.

Spain

Spain has developed at least two types of artillery rocket:

D-10—has ten rockets arranged in two rows of five.

E-21—has 21 rockets arranged in three rows of seven.

Both of the above systems are mounted on the rear of a Barreiros Panter 111 truck

The Spanish E-21 has 21 rockets on a 6×6 chassis

with the rockets pointing to the rear. There was also a version of the E-20 that had 20 rockets mounted on a two-wheeled trailer mount. No performance details have been made public.

Spanish rockets. On the left trucks are E-21s and on the right trucks are D-10 rockets.

80mm DIRA Rockets Switzerland

These have been developed and built by Oerlikon-Buhrle. The DIRA rocket is an unguided spin and fin-stabilised rocket and is fired from the Oerlikon multiple rocket launcher. For example the M-113 can be fitted with a 28 barreled launcher on the roof. The rockets have a calibre of 81mm, a muzzle velocity of 490 m/s, range of 8700m, length of 1300mm, total weight is 16kg with the warhead weighing 7kg.

130mm Multiple Rocket Launcher M-63 Yugoslavia

This is very similar to the American M-21 multiple rocket launcher and may well have been developed from that weapon. The M-63 consists of four rows each of eight barrels, making a total of 32 barrels. It fires HE rockets (weight 24·2kg and length 800mm) to a maximum range of 8200m. The 32 rounds can be fired in 12seconds. The M-63 has artillery carriage type trails and the carriage is provided with elevating and traversing mechanisms, large tyres are provided enabling it to be towed behind a truck. It has a crew of eight men.

The Yugoslav 130mm Multiple Rocket Launcher M-63

Launcher, Rocket: Multiple, 115mm, M-91
United States

The M-91 (development designation T-145) was designed to fire the M-55 chemical rocket with either a VX or GB warhead. In fact, apart from the training rocket (M-61) and dummy rocket, it could only fire the M-55. It has a total of 45 tubes arranged in five rows of nine each, the 45 rockets could be ripple fired in some 15 seconds. Emplacement time, including loading the 45 rockets was 30 minutes.

The M-91 is transported (i.e. carried in) a 6×6 M-35 truck, the small wheels of the M-91 being designed for moving the launcher only short distances. The mounting is provided with jacks for levelling purposes. The M-91 could also be fired whilst still in the truck and was air-transportable. Brief data is:

Weight of Launcher (without rockets): 545kg
Weight of Launcher (with rockets): 1914kg
Length: 3861mm
Width: 2972mm
Height: 1702mm
Traverse (on wheels): 360°
Traverse (left or right): 10°
Elevation: +1° to +60°
Range: 3000m–12,000m

Employment
The present status of the M-91 is uncertain. Some years ago three M-91s were assigned to each divisional direct support battalion. Also as it fired a chemical round (each rocket could cover an area of 2·59square kilometres) it has probably been phased out of US Army and Marine Corps service. None were exported to other countries.

The M-91 115mm Multiple Rocket Launcher being fired

MARS
United States

In October 1969 it was reported that the United States Army was developing MARS—Multiple Launched Artillery Rocket System. At that time it was in concept formulation. MARS will be an indirect fire system and deployed against such targets as tanks and armoured personnel carriers.

145mm Multiple Rocket Launcher M-21
United States

This was developed during World War II (development designation was T-66). The M-21 has a total of 24 tubes arranged in three rows of eight each. Weight of the launcher is about 694kg and range of the HE or smoke rockets (weight is 17·7kg) is 2200m to 8240m. The M-21s carriage has trails like a conventional artillery carriage, this enables it to be towed by a 4×4 truck, with another truck carrying the ammunition.

Employment
No longer in front line service with the United States but may be employed in other countries.

Soviet and Satellite Multi-Rocket Launchers

Designation	BM-21	BM-13	M-1965	BM-14-16	BM-14-17
Calibre	122mm	132mm	140mm	140mm	140mm
Number of Rounds	40	16	16	16	17
Type of Launcher	Tube	Rail	Tube	Tube	Tube
Mount-Vehicle	URAL-375	ZIL-151	—	ZIL-151 ZIL-157	GAZ-63
Chassis Type	Truck	Truck	Carriage	Truck	Truck
Rocket Length	2740mm	1473mm	1092mm	1092mm	1092mm
Rocket Weight kg	45·9	42·5	39·6	39·6	39·6
Rocket Range m	15000+	9000	10600	10600	10600
Rocket m/v m/s	—	350	400	400	400
Elevation	0° +50°	+14° +45°	0° +45°	0° +50°	0° +45°
Traverse	240°	20°	30°	140°	210°
Time to Reload (min)	10	5–10	4	3–4	3–4
Crew	6	6	5	7	6
Country	USSR	USSR	USSR	USSR	USSR

Full Designation/Notes/Employment

122mm Rocket Launcher (40 round) BM-21
Also called BM-21/40, first seen in Moscow in 1964, used by WPF including East Germany, Hungary, Poland, Soviet Union.

132mm Rocket Launcher (16 round) BM-13 (also known as M-13)
Used by WPF for training, for example East Germany and Poland.

140mm Rocket Launcher (16 round) M-1965
First seen in 1965 mounted on a carriage with split trails, used by airborne units.

140mm Rocket Launcher (16 round) BM-14-16
Entered service in 1954. Used by Algeria, China, Czechoslovakia, East Germany, Egypt, North Vietnam, Poland, Soviet Union, Syria.

140mm Rocket Launcher (17 round) BM-14-17
First seen in 1959 in service with Poland and the Soviet Union and other WPF.

200mm Rocket Launcher (4 round) BMD-20
Only used by North Korea. May be used for training by WPF.

240mm Rocket Launcher (12 round) BM-24
Used by Algeria, Egypt, Syria and WPF including East Germany, Poland and the Soviet Union.

240mm Rocket Launcher (12 round) BM-24 on AT-S
Used by the Soviet Union.

250mm Rocket Launcher (6 round) BM-25
Entered service in 1957, also called BMD-25, used by Soviet Union.

130mm Rocket Launcher (32 round) M-51
Also known as the RM-130. Used by Austria, Bulgaria, China, Czechoslovakia, Egypt, Romania.

140mm Rocket Launcher (8 round) WP-8
This was first seen in 1963 and is towed by a GAZ-69, 4×4, light truck. Used by Polish airborne units.

130mm Rocket Launcher (32 round) M-51 on ZIL-151/ZIL-157
Romania is the only country to use this combination, the normal mount for the M-51 being the Praga V3S truck.

BMD-20	**BM-24**	**BM-24**	**BM-25**	**M-51**	**WP-8**	**M-51**
200mm	240mm	240mm	250mm	130mm	140mm	130mm
4	12	12	6	32	8	32
Frame	Frame	Tube	Frame	Tube	Tube	Tube
ZIL-151 ZIL-157	ZIL-151 ZIL-157	AT-S	KrAZ-214	PRAGA V3S	—	ZIL-151 ZIL-157
Truck	Truck	Tractor	Truck	Truck	Carriage	Truck
3110mm	1180mm	1180mm	5822mm	800mm	1092mm	800mm
194	112·5	112·5	455	24·2	39·6	24·2
20,000	11,000	11,000	30,000+	8,200+	10,600	8,200
—	—	—	—	410	400	410
+9° +60°	0° +65°	0° +45°	0° +45°	0° +50°	0° +45°	0° +50°
20°	140°	210°	6°	240°	30°	240°
6–10	3–4	3–4		2	2	2
6	6	6		6	5	6
USSR	USSR	USSR	USSR	Czechoslovakia	Poland	Romania

ABOVE
Rear view of the BM-21 (40 round) Rocket Launcher on URAL-375 truck

BELOW
122mm Rocket Launcher (40 round) BM-21 being prepared for firing

RIGHT
140mm Rocket Launcher (16 round) M-1965 in action

BELOW
240mm Rocket Launcher (12 round) BM-24. Note the loading tray for loading the rockets in position at the rear.

FOOT
Czechoslovakian 130mm Rocket Launcher (32 round) M-51

Comparative Data on Self-Propelled Guns

Name	Sexton	AMX-105	AMX-155 F3	Abbot	JPZ 4-5
Country	Canada	France	France	Britain	Germany
Calibre	88mm	105mm	155mm	105mm	90mm
Type (Role)	Gun	Howitzer	Howitzer	Gun	Anti-Tank
Elevation	−9° +40°	−4½° +70°	0° +67°	−5° +70°	−8° +15°
Traverse	25° left 15° right	20° left 20° right	Note 3	360°	15° left 15° right
Rate of Fire	Note 1	6rds 20 sec	3rpm max 1rpm sust	12rpm	12rpm
Maximum Range	Note 1	15,000m	20,000m	17,000m	2000m
Ammunition Carried	112	56	nil	40	51
Reference		Note 2	Note 3	Note 4	Note 5

Note 1
For full details see the section on the 25 Pounder.

Note 2
There are two types of barrel: 30 cal length used by Netherlands; 23cal length used by France.

Range 15,000m with Mk 63 French round, hollow base, weight of shell 13kg, m/v 675 m/s (30 cal barrel). Range 11,500m with HEMI ammunition. Shaped charge round, 700 m/s will penetrate 150mm of armour at angle of incidence of 65°, or 360mm at angle of incidence of 0°. The 56 rounds carried included six anti-tank rounds.

There is also a version mounting the 105mm gun in a turret with a traverse of 360°, this however has so far been built in prototype form only.

Note 3
Traverse is 20° left and 30° right (gun at 0° to +50°).

Traverse is 15° left and 30° right (gun at +50° +67°).

Ammunition is carried in a AMX VCA tracked vehicle.

Ammunition:
French 155mm Model 56 HE shell, range 20,000m, projectile weight 43·75kg.
American M-107 shell range 18,000m, projectile weight 43·10kg.
French TA 64 hollow base shell, range 21,500m, projectile weight 42·80kg, muzzle velocity 570/740 m/s depending on charge.

Also there is a French smoke round weight 44·25kg and an illuminating round weight 44kg.

Under development is a 155mm Brandt round with additional propulsion, this will have a range of some 35,300m.

Note 4
There is also an anti-aircraft vehicle based on the Abbot chassis called the Falcon, this was developed by Vickers and is armed with 2×30mm guns, it exists in prototype form only.

Of the 40 rounds carried six are HESH.

Rounds available are HE, Smoke, Target Indicating, Illuminating, HESH and Squash Head Practice. There are a total of eight charges.

Note 5
Full designation is Kanonenjagdpanzer. The gun is the same as that mounted in the M-47 and fires HEAT-T and HEP-T rounds.

Other self-propelled guns include:

French 155mm GCT SPG

This is under development and is based on the AMX-30 chassis. The gun is a 155mm, 30 cal in length with a traverse of 360°, elevation is −5° to +66°. It is fitted with an automatic loader and carries 42 rounds of ammunition. Sustained rate of fire will be 8 rounds a minute. Maximum range with the CCr TA 68 shell will be 23,500m.

Swedish 155mm Bandkanon 1A

This is in service with the Swedish Army and uses components of the S tank. The 155mm gun has a total traverse of 30°, elevation being −3° +40°. Barrel is cal 50. Magazine holds 14 rounds and range is 25km.

Swiss 155mm Panzer-Kanone 68

This is based on the Pz.68 tank chassis and the 155mm gun has a range in access of 30km.

United States Self-Propelled Gun Characteristics

M-52 and M-52A1 Self-Propelled Howitzer

Armed with a 105mm M-49 gun in mount M-85, elevation −10° to +65°, traverse 60° left and right. Breechblock is of the vertical sliding type. Rate of fire is 3rpm. Ammunition used includes HE (472 m/s), HEP, HEP-T, HEAT (m/v 380 m/s HEAT-T, chemical and illuminating. HEAT round has a range of 6130m and HE 11,150m. 105 rounds are carried.

M-108 Self-Propelled Howitzer

Armed with a 105mm gun with an elevation −6 to +75°, traverse 360°. Breechblock is of the sliding wedge-trap type. Rate of fire is 3rpm. Ammunition used includes HE (range 12,000m), illuminating, leaflet, smoke and gas. 87 rounds are carried.

M-53 Self-Propelled Gun

Armed with a 155mm gun M-46 in mount M-86, elevation −5° to +65°, traverse 30° left and right. Breechblock is of the stepped-thread, interrupted-screw, horizontal-swing type. Rate of fire is 1rpm. Ammunition used includes HE (854 m/s, range 23,500m), chemical, smoke and illuminating. 20 rounds of ammunition are carried.

M-44 and M-44A1 Self-Propelled Howitzer

Armed with a 155mm gun M-45 in mount M-80, elevation −5° to +65°, traverse is 30° left and right. Breechblock is of the interrupted stepped-thread type. Rate of fire is 1rpm. Ammunition used includes HE (m/v 569 m/s, range 14,600m), chemical, illuminating and smoke. 24 rounds of ammunition are carried.

M-109 and M-109A1 Self-Propelled Howitzer

The M-109 is armed with a 155mm gun M-126 in mount M-127, elevation −5° to +75°, traverse 360°. Breechblock is of the welin-thread interrupted screw type. Rate of fire 1rpm. Ammunition used includes HE (m/v 563 m/s, range 16,800m) canister, chemical, nuclear, smoke, illuminating. The M-109A1 has a longer barrel and fires a HE round to a range of 18,000m, m/v being 635 m/s, US Army M-109s were being modified into M-109A1s starting in 1972. There is also a modified German (M-109G) and Swiss (M-109U) version. M-109 carries 28 rounds of ammunition.

M-107 175mm Self-Propelled Gun

Armed with a 175mm gun M-113 in mount M-158, elevation 0° to +65°, traverse 30° left and right. Breechblock is of the stepped-thread/interrupted screw type. Rate of fire one round every minute. Fires a HE or chemical round to a range of 32,800m, HE round has a m/v of 923 m/s.

M-55 Self-Propelled Howitzer

Armed with an 8in gun M-47 in mount M-86, elevation −5° to +65°, traverse 30° left and right. Breechblock is of the stepped-thread, interrupted-screw, horizontal-swing type. Rate of fire is one round per minute. It fires a HE round with a m/v of 594 m/s to a range of 16,925m, it can also fire a nuclear round. A total of 10 rounds are carried.

M-110 Self-Propelled Howitzer

Armed with an 8in gun M-2A1E1 in mount M-158, elevation 0° to +65°, traverse 30° left and right. Breechblock is of the interrupted-screw, stepped-thread type. Fires a HE (m/v 594 m/s) or nuclear round to a maximum range of 16,930m. Maximum rate of fire is one round per minute.

The M-110 and the M-107 will be replaced in the United States Army by the M-110E2 8in Howitzer, this has a barrel some 2·44m longer than the standard M-110 and will have a much greater range. Ammunition being developed for the M-110E2 includes HE, incendiary, nuclear, improved conventional munitions (ICM) and dual purpose (DP).

Other United States self-propelled artillery pieces still in service with various armies (not the United States) include the M-56 Self-Propelled Anti-Tank Gun, M-10, M-10A1, M-36, M-36B1, M-36B2, M-18 Tank Destroyers, and the M-7 and M-7B1 Self-Propelled Howitzer. Full details of these and of the vehicles listed above will be found in *Armoured Fighting Vehicles of the World*.

Soviet Self-Propelled Gun Characteristics

ASU-57 (Airborne Assault Gun)

57mm gun with an elevation of −5° +20°, total traverse of 12°, 30 rounds carried. HE projectile wt 2·8kg, m/v 706 m/s, APHE wt 3·1kg, m/v 990 m/s, HVAP wt 1·8kg, m/v 1255 m/s. Rate of fire 6–10rpm. APHE round will penetrate 106mm at 500m and HVAP round 140mm of armour at 500m.

ASU-85 (Airborne Assault Gun)

85mm gun with an elevation of −4° +15°, total traverse of 12°, 40 rounds carried. HE projectile wt 9·5kg, m/v 792 m/s, APHE wt 9·3kg, m/v 792 m/s, HVAP wt 5·0kg, m/v 1030 m/s. Rate of fire 3–5rpm. APHE round will penetrate 102mm of armour at 1000m, HVAP round 130mm of armour at 1000m.

SU-76 (Assault Gun)

76·2mm M-1942/1943 with an elevation of −3° +25°, total traverse of 32°, 60 rounds of ammunition carried. HE projectile weight 6·2kg, m/v 680 m/s, APHE weight 6·5kg m/v 655 m/s, HVAP wt 3·1kg, m/v 965 m/s, HEAT wt 4·0kg, m/v 325 m/s. Rate of fire 15rpm. APHE round will penetrate 69mm at 500m or 60mm at 1000m, HVAP 92mm at 500m or 58mm at 1000m, HEAT round 120mm at 500m or 1000m. Only used initially as an assault gun.

SU-100 (Assault Gun)

100mm M-1944 (D-10S) gun with an elevation of −2° +17°, total traverse 17°, 34 rounds of ammunition carried. HE projectile wt 15·7kg, m/v 900 m/s, APHE wt 15·9kg, m/v 1000 m/s. Rate of fire 7rpm. APHE round will penetrate 185mm at 500m or 1000m, HEAT round 380mm at 500m or 1000m.

ISU-122 (Assault Gun)

122mm M-1944 (D-25S) with an elevation of −3° +16°, total traverse 14°, 30 rounds of ammunition carried. HE projectile wt 25·5kg, m/v 800 m/s, APHE wt 25kg, m/v 800 m/s. 3 rounds per minute. APHE round will penetrate 190mm at 1000m. Other model of ISU-122 has the A-19S gun, this has an elevation of −4° +15°, total traverse 11°.

ISU-152 (Assault Gun)

152mm M-1937/44 (ML-20S) with an elevation of −3° +20°, total traverse 10°. Only 20 rounds of ammunition carried. HE projectile wt 43·6kg, m/v 655 m/s, APHE wt 48·8kg, m/v 600 m/s. Rate of fire 2–3rpm. APHE round would penetrate 124mm of armour at 1000m.

ZSU-23-4 (Self-Propelled Anti-Aircraft Gun)

Armed with four turret mounted 23mm guns with an elevation of −5° +80°, with a total traverse of 360°. Fires HEI and API rounds, these both have a projectile weight of 0·19kg and a muzzle velocity of 970 m/s. Cyclic rate of fire is 800/1000 rounds per barrel per minute, practical however is 200rpm. The API round will penetrate 25mm of armour at 500m. Effective anti-aircraft range is 2000m. This weapon has a radar fire control system mounted to the rear of the turret, this folds down behind the turret when not required.

ZSU-57-2 (Self-Propelled Anti-Aircraft Gun)

Armed with twin 57mm S-68 guns with an elevation of −5° +85°, total traverse is 360°. Fires HE and APHE rounds. HE projectile wt 2·8kg, m/v 1000 m/s, APHE projectile wt 3·1kg, m/v 1000 m/s, rate of fire is 70rpm, per barrel. Cyclic rate of fire is 105–120rpm. The APHE round will penetrate 106mm of armour at 500m. Effective anti-aircraft range is 4000m. No fire control system is fitted, only the sights on the gun.

NOTE: *Full characteristics of the above vehicles is given in the book Armoured Fighting Vehicles of the World.*

Comparative Data on Self-Propelled Anti-Aircraft Vehicles

Name	M-53/59	AMX-30 DCA	AML S-530	Leopard	M-42	M-163 (Vulcan)
Country	Czech.	France	France	Germany	USA	USA
Calibre	30mm	30mm	20mm	35mm	40mm	20mm
Elevation	$-10° +90°$	$-8° +85°$	$-10° +75°$	$-5° +85°$	$-5° +87°$	$-5° +80°$
Traverse	360°	360°	360°	360°	360°	360°
Rate of Fire	450/500*	650*	740*	550*	240*	1000/3000**
M/V HE	1000m/s	Note 1	Note 2	Note 3	Note 4	Note 5
M/V API	1000m/s					
Effective Range A/A	2000m	3000m	2500m	4000m	4600m	Note 5
Ammunition Carried	N/A	1200	600	700	480	1200+800**
Radar	No	Yes	No	Yes	No	Note 5

* cyclic rate of fire, per barrel
** 1000rpm ground role, 3000rpm A/A role, the 800 rounds are in reserve

Note 1
Rounds available are:

EP	inert practice shell w/o tracer	shell wt 360gr	wt of cartridge 870gr
ET	inert practice shell with tracer	shell wt 360gr	wt of cartridge 870gr
UIA	HE incendiary shell	shell wt 360gr	wt of cartridge 870gr
UIAT	HE incendiary shell with tracer	shell wt 360gr	wt of cartridge 870gr
RT	AP shell with tracer	shell wt 390gr	wt of cartridge 900gr
RI	AP incendiary shell w/o tracer	shell wt 390gr	wt of cartridge 900gr
RIA	AP incendiary shell, self destruct	shell wt 360gr	wt of cartridge 870gr
RINT	AP incendiary shell+hard core and tracer	shell wt 360gr	wt of cartridge 870gr

Maximum traversing speed=80° a second.
Maximum elevating speed=40° a second.
Maximum A/A range is 10,000m.
Can fire five or 15 round bursts.
Note also that the AMX-13 tank with 2×30mm guns has the same turret and guns as the AMX-30 DCA.

Note 2
Rounds available are:
Tracer piercing shell, m/v 1000 m/s, armour penetration 23mm at 1000m.
Tracer explosive shell, m/v 990 m/s.
Explosive shell m/v 1020 m/s. There is also an incendiary round.
The guns fitted are the AME 621, these can fire single rounds, bursts or continuous firing.
Maximum traversing speed=80° a second.
Maximum elevating speed=40° a second.

Note 3
Rounds available include:
APHE wt 1562gr m/v 1175 m/s.
HEAT wt 1175gr m/v 1175 m/s, will penetrate 35mm of armour at 1000m.
APT wt 1562gr m/v 1200 m/s, will penetrate 66mm of armour at 1000m.
The total of 700 rounds of ammunition includes 660 A/A rounds and 40 for use against ground targets (i.e. APT or HEAT) The guns can fire single shots, bursts, or a continuous burst. The German model is the 5PZF-B and the Dutch the 5PZF-C.

Note 4
Fires single or bursts, ammunition used is:
AP-T, m/v 870 m/s, maximum range in ground use is 8655m.
HE-T, m/v 870 m/s, maximum range in ground role is 4760m.

Note 5
For details of the ammunition etc., see the section on the towed version of the system, the M-167. The system has radar for ranging purposes only.

155mm gun at point of firing. The antenna unit is to the left of the wheels.

Doppler Radar DR 513 — Denmark

Muzzle Velocity Measurement Equipment

The DR 513 is built by Radartronic A/S and is based on the doppler principle. It can measure muzzle velocities ranging from 50 to 2000 m/s, accurate to 0·1%.

The basic equipment for measuring the muzzle velocity consists of: **A.** The antenna unit. **B.** The data unit. **C.** The test unit. The doppler signal from the antenna unit is directly proportional to the velocity and is fed to an advanced electronic data unit where the results can be read out on four display tubes. The figure on the display is the time the projectile has spent travelling a distance of 2m measured 30 and 40metres in front of the muzzle. The corresponding velocity can easily be found in tables. The correction factor for the geometrical error introduced by offsetting the antenna from the line of fire can also be found in the tables. The test unit is a simulator providing a fast control on function and transmitter frequency.

The antenna unit is designed to withstand a very high degree of shock and vibration and can be placed next to or even on the gun, to provide a minimum geometrical error. The use of the DR 513 makes it possible to measure the velocity of all sizes of projectiles. The measuring range is about 5000 times the calibre.

The DR 513 basic equipment can be extended with various units for analog and digital measurements throughout the trajectory. The analog data unit together with an ultra violet paper recorder give a curve showing the velocity as a function of time. The ballistic trajectory analyser (BATRAN) system together with a tape puncher give results on a punched tape ready for data handling in a computer. The results can also be printed directly on a digital printer.

Employment

Danish Forces, Finland, Germany, Great Britain, Holland, Italy, Malaysia, Mexico, Norway, Pakistan, Singapore, Sweden, Switzerland, Yugoslavia.

Eldorado

Anti-Aircraft Fire Control System

The Eldorado (TH.D.1229) was jointly designed and developed by the Compagnie Française Thomson Houston (radar and servo-mechanisms) and Officine Galileo of Florence (trailer, computer and optics). The system consists of the radar and its reception and display elements, optical system with simultaneous display of the radar screen and optical picture (has simultaneous display of the radar screen and optical picture), the computer, operators desk (only one man is required to operate the system), power supply and air conditioning unit.

The trailer weighs only 1600kg, and also contains the circuitry and display equipment for the Mirador (TH.D.1350) acquisition radar, it can also transport the Mirador antenna inside the trailer and its tripod outside the trailer. Other types of radar could be used in place of Mirador.

When Eldorado is used without an external acquisition system is can perform the acquisition on its own by:
Helical scanning, 360° in azimuth; in elevation the scanning is automatic from 1·5° to 13·5°.
Sector scanning on elevation and azimuth; scanning angle 10°.

The Mirador consists of the aerial and tripod weighing a total of 52kg, transmission unit weighing 90kg, receiving unit weighing 65kg and the PPI indicator which weighs 25kg and is housed in the trailer. Its maximum range is some 18km, the antenna rotates at 60rpm and its maximum detection altitude is 2500m.

Employment
France.

Other Thomson Houston Radars include the Cotal for 90mm A/A guns, the Super Cotal which was a development of the Cotal and was for use on firing ranges, the TCP-TPO and TPM for 40mm Anti-Aircraft Guns and the TVR 30, this was a light fire control equipment for 30mm Anti-Aircraft Guns and had television tracking equipment.

TOP
Eldorado trailer

ABOVE
Mirador in operation

AMETS

Artillery Meteorological System

This has been developed by Marconi Space and Defence Systems Limited and the Plessey Company Limited. Plessey developed the radio-sonde and the WF3M radar, the latter being developed from the civil WF3.AMETS is a self-contained mobile meteorological system designed for obtaining and processing information on upper air conditions.

It consists of a Command Post Vehicle (CPV), this tows the generator, $\frac{1}{4}$ton reconnaissance vehicle, instrumentation vehicle (AIV), this tows the radar trailer, optical tracker, stores vehicles, hydrogen trailers, balloon filling shelter and meteorological balloons. These balloons consist of a radar reflector, radio-sonde, paper parachute and the balloon itself.

Basically a balloon is sent aloft, to this is attached the radar reflector and the radio-sonde transmitter, the balloon is automatically tracked by radar, while the radio-sonde transmits a signal conveying temperature. This signal is passed through a special converter which transforms the received radio-sonde signals into a form suitable for input into the computer. Data from the balloon tracking radar is also fed into the computer. Pressure is derived by the computer from radar height and surface pressure using standard ICAO atmosphere figures corrected for temperature. Average humidity values at each altitude are available on punched tape for all areas where AMETS operates. These values are fed into the computer store using the teleprinter. The computer used is the 920B, this is the same as that used in FACE.

The computer can provide eight types of meteorological messages i.e. standard artillery computer message (i.e. fed directly into FACE), standard ballistic message, surveillance drone message, sound ranging message, weapon locating radar message, biological/chemical message, nuclear fallout message and civil forecasting message. In addition ten partial messages can also be produced during a sounding.

The CPV is the communications and control centre and contains the output teleprinter, the rear half of the vehicle also carries spare radio-sondes. The AIV contains the computer, radar display, teleprinter, digital display, sonde monitor, programme loading unit, DPE console, computer test set, power supply test set and stand by batteries.

Both the AIV and the CPV are pallet mounted on 4ton 4×4 trucks and are air-conditioned and fitted with NBC equipment. AMETS can operate in temperatures ranging from −32° to +52°C, and up to 20,000m.

Employment

Operational with the British Army.

This is the CPV (Command Post Vehicle), and is the control and communications centre of AMETS

AWDATS

Great Britain

Artillery Weapon Data Transmission System

This has been developed by Marconi Space and Defence Systems Limited. Each AWDATS consists of a digital coding unit situated at the command post, and a data display at each gun. Gun firing data from FACE is transmitted via the coding unit over existing radio or line equipment to the display units. This takes $2\frac{1}{2}$seconds for a six gun battery. AWDATS is on order for the British Army.

Cymbeline

Great Britain

Mortar Locating Radar

Cymbeline (Radar Field Artillery No 15 Mk 1) has been developed by EMI Electronics Limited, in association with government establishments since the 1960s as a replacement for Green Archer.

Cymbeline is designed not only for mortar location duties but also for coastal surveillance, helicopter and light aircraft control, rapid survey, meteorological balloon tracing and the adjustment of fire by air bursts and ground bursts.

Its main features are its light weight, high accuracy of location and adjustment, high tactical mobility, high reliability, easy to operate and maintain, built in self-test facilities, built in target simulator for training purposes and it is very quick into action.

Cymbeline is similar in operation to Green Archer, it does however have a third radar beam elevation angle that may be used to alert the operator. The radar unit itself consists of three basic groups of assemblies—1 The mounting, 2 The radar head and 3 The remote control and indicator units, these can be operated up to 15m away from the radar. Unlike the Green Archer, Cymbeline has its own built in generator set, this is inaudible beyond 200m and it is powered by a Wankel engine. It is provided with 4 manual screw jacks for levelling purposes.

Its maximum displayed range is 20km, minimum detection range is 1km, maximum detection range depends on the calibre of the mortar, location time is 15–20seconds. Total weight of the radar and the trailer is 980kg. Normal operating crew is two men, although it can be operated by one man.

The display unit contains a built-in radar simulator which provides realistic synthetic signals to the cathode ray tube representing intercepts of mortar bombs. All main units contain built in test circuits by means of which the radar operator can locate a defective unit, this can then be quickly replaced. A data memory storage unit is available as an optional extra, this allows video signals from the receiver to be stored and displayed on the cathode ray tube.

Cymbeline in travelling position with the Radar reflector folded down

Cymbeline can be transported in four ways. First by mounting on a two-wheeled one tonne trailer towed by a Long Wheel Base Land Rover, second carried by a helicopter (i.e. a Whirlwind), third it can be quickly detached from the trailer and carried by four men using lifting poles and slings, 4th mounted on the roof of the FV 432 APC (or similar vehicle). In the latter case it is fitted with an automatic hydraulic levelling system.

Employment

In service with the British Army from 1973.

Ferranti Pacer

Muzzle Velocity Measuring Equipment

The Pacer has been developed by the Electronic and Display Equipment Division of Ferranti, and is a development of their Projectile Velocity Measuring Equipment which is used at British Ministry of Defence Proving Ranges.

The Pacer has been designed specifically for use by forces in the field. After setting up the equipment all the operator is required to do is to set one control for the approximate anticipated muzzle velocity and press a 'reset' button. The operation of the equipment is then automatically initiated by detection of the instant of fire by the muzzle flash detector. Within two seconds of firing, the achieved instrumental muzzle velocity is presented on a legible decimal display in metres/second units. The radar transmitter is activated at the instant of fire and is switched off automatically when the velocity measurement is completed. This is useful when enemy sensors are in the area.

The equipment incorporates circuitry which automatically checks the conditions under which each measurement is made are satisfactory. If these checks indicate that the measurement may be invalid an all-zero result is indicated. It also incorporates an overall test facility. The equipment will operate either from a 24 volt vehicle supply or from mains power. Pacer will operate in temperatures −20° to +55°C.

Pacer is accurate to ±0·30 metres/second for velocities between 200 and 1599 m/s provided that:

a the anticipated velocity control is set within ±20 m/s of the true velocity.
b The aerial is sited correctly 2 metres ±0·5 metres from the barrel axis between breech and muzzle and aimed along the line of fire with an error less than ±4° in elevation and ±8° in bearing. The built in clinometer gives adequate elevation accuracy.

The four components of Pacer are:

1 Flash detector, weight 1·50kg.
2 Aerial, weight 40kg.
3 Display unit, weight 34kg.
4 Cable and drum, weight 45kg.

Employment

The Pacer is fully developed and at least two overseas countries have purchased models for evaluation.

Pacer. On the left is the aerial (erected) and on the right the display unit ready for use. Cables are not connected in the photograph.

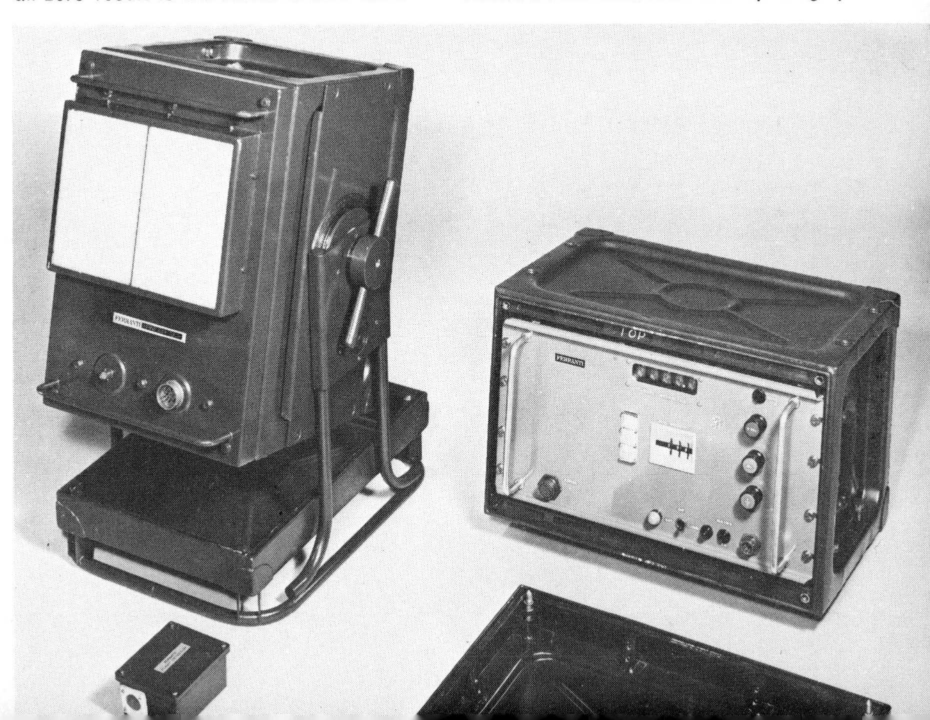

Field Artillery Computer Equipment (FACE)

Great Britain

The FACE system is produced by the Military Division of the Marconi Space and Defence Systems Limited and has been in service with the British Army since 1969. Its basic task is to automate the complex procedures associated with the preparation of gun-firing data for field artillery. Its three main advantages are: 1 Cut in training time. 2 Weapon response time reduced. 3 Greater accuracy. FACE can be fitted into a variety of vehicles including the FV 432, FV 610 and Land Rover. It can also be used in conjunction with AMETS.

FACE can be used in two roles:

Artillery Role

In this role every piece of data that is required to place the shell on the target is displayed on the console, this data includes the gun and target position, allowances for the rotation of the earth, state of the weather, wear of gun barrels, variations in projectile weight and many other pieces of data. Full computation takes only a few seconds.

Survey Role

In this role more than 34 problems are stored each comprising a routine for input, the calculation, and an output routine for presenting the answer on the teleprinter. The calculations take less than 10seconds. The system is based on the Elliott MCS 920B digital computer and uses a specially designed console for injecting the target and meteorological information, the console has four sections—data display, matrix, master controls and keyboard.

The complete system consists of the computer, console, power supply (and fan), supply control unit (and battery), inverter (power, static), line adapter, teleprinter and programme loading unit. It is silent and vibration free. The total weight of the system is some 300kg, power supply is 24 V, 150W (stand by) and 300W (operational)

A Computer Field Test Set has been developed to enable unskilled operators to quickly locate faults. Once located the logic boards or sub-assemblies can be quickly replaced.

Employment

British Army, one FACE system is deployed with each battery. It is also on order for a number of overseas armies.

FACE installation in a FV 432. On the left is the teleprinter and on the right is the console.

Green Archer Mortar Locating Radar

Great Britain

The Green Archer system was developed by EMI Limited in association with Government establishments from the mid-1950s. The first prototype was built in 1958 and the system entered service with the British Army in 1962. The role of Green Archer is the location of enemy mortar positions but it can also be used for the adjustment of artillery fire, surveillance duties and the control of light aircraft and helicopters.

The basic idea is that the radar operator plots two points on the mortar bomb trajectory after it leaves the weapon. The slant range and bearing to each of the plotted positions is measured, in addition the flight time between the two points is measured. The position of the mortar is then determined by the analogue computer from this information plus the pre-set elevation angles of the radar. It takes some 30seconds to determine the position of the enemy mortar. Maximum range is reported to be 17,000m.

Green Archer is employed in two versions, one trailer mounted and the other mounted on a tracked vehicle. They have the designation on the British Army of:

Radar Field Artillery No 8 Mk 1

This is the trailer mounted version. The radar is mounted on a four-wheeled trailer and there is a separate trailer for the silenced generator. Basic data of the radar trailer is: weight 2915kg, length of trailer 4m, length of trailer including towbar 6m, width 2m, height travelling 2m, height operational 2·9m and ground clearance is ·38m. Both the radar and generator trailers are towed by a 1ton armoured truck.

Radar Field Artillery No 8 Mk 11

In this version the Green Archer Radar is mounted on a modified FV 432 chassis, the rear being modified, this is then designated the FV 436, the vehicle retains its amphibious capability. The vehicle carries the crew and is self-contained. The German and Danish Armies have their Green Archer Radars mounted on their M-113s in a similar manner.

The following training aids have also been built:

Simulator, Radar, Jamming, No 2 Mk 1
Simulator, Radar, Target, No 3 Mk 1
Simulator, Radar, Target, No 4 Mk 1

There are 13 major sub-assemblies in Green Archer, and the fault can be quickly

The Green Archer Radar Trailer

The silenced generator used with the Green Archer

detected and replaced by the spares vehicle, this vehicle carries a complete set of spare parts and sub-assemblies. There is one of these vehicles in each Green Archer detachment.

Employment

Denmark, Great Britain, Italy, Israel, South Africa, Sweden, Switzerland. In the British Army there is one Green Archer detachment per close support artillery regiment, this consists of two radars, one command post and a spares vehicle.

L 4/5 Weapon Control System Netherlands

This system was developed and built by N. V. Hollandse Signaalapparaten of Hengelo in the late 1950s and early 1960s. Extensive trials were carried out in 1962 and 1963 and these proved that the system was an effective answer to low level attack, it can be used with both anti-aircraft guns (40mm) or guided missiles. Normally one L 4/5 controls three 40mm guns. The complete system is all on one trailer, although another trailer is used for the generator. The six main components of the system are the search antennas, tracking antenna, monocular sight, operators' cabin, velocity measuring equipment and the converter.

There are two search antennae back to back, one of these is for medium and low targets and the other for very low targets, one tracking antenna and a monocular sight for optical interception and location. The control panel consists of the panel with a PPI for medium and low targets and a PPI for very low targets, an A-scope for hand tracking, testing and operational controls and the control panel. The digital transistorized computer is under the cubicle.

Its main features are the detection of aircraft at very low to medium heights, extensive anti-jamming features, IFF facilities, short reaction time, rapid search scan, automatic picking-up and tracking of targets, search-while-track through 360°, digital computer programmed for the simultaneous control of three guns, possibility of automatic connection with a central position by means of digital data links, simple operation, high degree of mobility.

FASCAN Search Pattern. The system is so designed that all targets within the search range are detected at an adequate range. The PPIs provide a clear survey of the tactical situation. The spiral scan pattern of the high beam can be tilted through about 3·4°. The axis of the low beam can be tilted between −0·8° and +3·4°. The beams thus overlap.

Modes of Operation—the search radars can be either continuous search or sector search. Picking-up, automatically by radar or manual, i.e. the director operator can pick up any target visually with the aid of two handles with built-in grip switches. Tracking—automatic or manual. In automatic it can be either auto-radar, generative or free generation. In manual it can be optical or on the PPI.

Technical data: Search radar maximum search range is 31km, maximum PPI range 34km. Tracking antenna—maximum tracking range is 32km. Maximum target speed horizontal 500 m/s, vertical 300 m/s. Crew is 2 or 3 men, weight 6000kg. Height operational is 5200m, height travelling 3300mm, length 7500mm, width 2200mm. And can operate in temperatures from −25°F to +110°F and in wind forces up to 75km/h.

Employment

Netherlands.

ABOVE
The L 4/5 all weather fire control system

BELOW
The L 4/5 all weather fire control system with three 40mm LAAGs in the background and the generator on the right

The new VL 4/41 (Flycatcher) fire control system

VL 4/41 (Flycatcher) Weapon Control System

Netherlands

This is one of the more recent developments by Signaal and is designed for point and area-defence against a hostile air attack. It can control up to three guns of the same type, or a mixed group of two guns and a guided weapon. The complete system is housed in a standard military container. When operational the radar antenna group is on top of the roof, when being transported it can be retracted into the container. The radar group consists of a separate search and tracking antenna rotating on the same axis. IFF dimpoles are installed in the housing of the search antenna, and a TV camera can be mounted on the tracking antenna.

The system is based on the system developed for the Leopard A/A tank. The VL 4/41 enables all types of aircraft and missiles within its range of 20km to be engaged after an automatic alarm and IFF interrogation. It has a SMR-S digital computer.

Employment
Trials.

Simrad Laser Range Finder PM 81 Norway

The Simrad Laser Range Finder is a one-man portable, electro-optical instrument designed for use by artillery observers. It contains in one unit the laser transmitter, receiver system, electronic circuits, sighting telescope and goniometer. Total weight is 6·4kg.

The instrument instantly measures range, azimuth and elevation angles to possible targets and may be used for determination of opposition, target position and for adjustment of fire. It can measure ranges from 200–20,000 metres with an accuracy of ±5 meters. The azimuth angle range is 6400 mils and elevation angle is ±350 mils, both with a resolution of 1 mil.

The range finder fits on a special lightweight tripod, designed for field and combat conditions. The tripod can be adjusted heightwise, and the range finder fits to it by means of a tripod screw.

The battery pack consists of NiCd cells and is clamped to the tripod, the battery has a rated voltage of 24V and its capacity is some 300 firings at +20°C. The battery cable is 80cm long.

A special sighting telescope with reticle is combined with the optical receiver. Under normal conditions the range finder is able to range any target within 6000 metres that will produce an image covering an open 1 mile wide area in the centre of the reticle. The reticle is illuminated by beta light for night use.

The 20° angle of the eyepiece allows the operator to wear a helmet when he aims the instrument through the sighting telescope. The top of the range finder housing has a quick-aiming sight which enables the range finder to be rapidly pointed against targets appearing only for a short time. The range finder housing is waterproof.

The operation of the instrument is based upon the echo principle. When the trigger is activated, a short, invisible light pulse of high intensity is transmitted from the laser optics to the exit window. When hitting a target, some of the transmitted light is reflected back towards the instrument, it passes through the entrance window and activates the receiver.

The range is displayed directly in metres and up to three individual targets (if located at least 50 metres apart) can be displayed simultaneously. The display also provides indication of targets within the minimum range, low battery voltage and low laser power. The display will light for 10seconds, then extinguish.

The minimum range sets the range within which reflections from targets shall be rejected. The minimum range may be set at any distance between 200 and 6000 metres. Total weight of the instrument including tripod is about 11kg.

An option is digital output of range and angles to external computers.

The PM 81 has been developed by Simonsen Radio A.S. of Norway for the Norwegian Army under a contract with the Army Material Command. The development was supported by the Norwegian Defence Research Establishment. Some have been ordered for the British Army.

The Simrad Laser Range Finder PM 81

The SAAB ACE-380 in operational use

SAAB ACE-380 (Artillery Computer Equipment)

Sweden

The ACE-380 is a Field Artillery Computer Equipment for firing data calculations and is built by Saab-Scania. The system can be integrated with existing fire control systems. It operates at battery level but can also supply firing data to neighbouring batteries.

Its main characteristics are: single unit design, portable by two men, simple one-man operation, general purpose digital computer (Datasaab D5-30), built in fault detection facilities, fixed or moving target data calculation, firing data within one second and easy training.

The use of a multi-purpose digital computer with ferrite core memory and real-time programming form the basis for a number of tactical and operational capabilities, from the point of view of battery staff nothing is new, only much faster.

Its capabilities include target data storage for 500 targets, firing data for up to 40 ballistics inclusive of corrections and daily adjustments, storing of complete target data with corrected firing information for 20 targets, automatic or semi-automatic input from forward observer, automatic output of firing data to gun data link, automatic input from transmitter of muzzle velocity data, prognosis of air for 30 levels above the ground. All data is displayed immediately when required and is easily updated.

The keyboard and the display panel are designed with separate functional fields for each group of parameter data input. All input data can be checked on the panel before executing orders are released.

The keyboard is divided up into the following operational fields: fire adjustment, changing of target, battery position data, target data, ballistic data and influences (i.e. air pressure, muzzle velocity, projectile weight etc.). There are also clear, store, compute, test, fire etc. buttons. In addition there is an alpha numerical key area.

Employment
Swedish Army.

Skyguard

Fire Control System

Skyguard has been developed by Contraves AG in co-operation with LM Ericsson of Sweden and Albiswerke Zurich AG. It is a mobile miniaturised all-weather fire control system for use against low-flying aircraft and air to surface missiles. It can be used to control anti-aircraft guns or guided missiles.

The system consists of the following:

Pulse Doppler Search Radar. This has high detection probability, high information rate, fine resolution cell, high clutter suppression, automatic target alarm, automated acquisition procedure for tracking radar, resistant to ECM, elaborate video presentation and can have IFF facilities.

Pulse Doppler Tracking Radar. Monopulse-evaluation, fine resolution, accurate tracking, automated target exchange under computer control, ASM detection and alarm, resistant to ECM and second tracking system when tracking missiles.

TV Tracking System. Automatic high-precision target tracking, using video-processor, computer aided tracking using a joystick Vidicon (with or without light intensifier) or SEC-Vidicon camera available. The tracking mount supports the tracking antenna and TV Camera as well as the independently rotating search radar.

Contraves Cora 11 M Digital Computer. This carries out the following tasks: threat evaluation based on search radar information, hitting point data for the guns, calculations for guided missile deployment, calculations of tactical target data, monitoring of all Skyguard sub-systems. Full automatic functional and performance check of the entire system. Simulation of combat situations for realistic training.

Digital-Data Transmission. Standard two-wire field telephone connection cable to the weapons.

Control Console. This can be operated by one or two men and has an indicator (PPI) displaying doppler and raw-video as well as symbols and markers for the presentation of the tactical situation. A scope is provided to check target tracking and to judge the ECM situation. Tactical display with numerical read-out, monitoring of the TV tracking system, rolling-ball to control markers, joystick for manual control of the tracker, matrix panel for data input and output, and an intercommunication system.

Power Supply System. This is mounted internally but can be removed for external operation, it consists of an air cooled, four stroke petrol engine.

The system is mounted in a body made of fire-resistant reinforced fibre-glass polyester and is fully air conditioned. The tracker folds away whilst travelling. It has automatic hydraulic fine levelling and has good accessibility for maintenance purposes. The body is air-transportable and can be mounted in the rear of various types of vehicle.

Employment

Has been tested by the Swiss Armed Forces.

Skyguard in transport position being towed by a 4×4 truck

BELOW
Skyguard mounted in the rear of a M-548 tracked vehicle, ready for action

FOOT
Skyguard ready for action

Super-Fledermaus

Anti-Aircraft Fire Control System

The Super-Fledermaus Anti-Aircraft Fire Control System has been developed by Contraves AG in association with Oerlikon Machine Tool Works (Bührle & Company) (who developed the muzzle velocity measuring equipment), and Albiswerk Zurich AG. Over 20 countries use this system.

The complete equipment is mounted on a four-wheeled cross-country trailer and consists of the following:

1 Tracker—with one man control, telescopic sight and radar antenna for target acquisition and tracking.
2 Electronic Computer—continuously determines the firing data for three individual gun emplacements.
3 Acquisition and Fire Control Radar—with turntable magnetron transmitters which can be switched over during action in case of jamming.
4 Muzzle Velocity Measuring Equipment—measures the muzzle velocity of the three guns individually.
5 Optical Putter-On—This enables the tracker to be directed immediately onto targets appearing unexpectedly.
6 Signal box—enables the fire control officer to give orders for target acquisition, tracking and opening fire by remote control.

The system can acquire targets up to 50km away by night or day and track them automatically from a range of 40km. It can do this by the following five modes of operation:

Helical scanning, the antenna rotates in bearing and slowly threads its way down. In this manner the entire horizon is scanned in a gapless helical sweep.

Sector scanning, a vertical oscillating motion is imparted to the antenna, at the same time the antenna is slowly rotated in bearing so that the radar beam scans a specified sector of space without a gap by rowing one path to the next.

Early-warning radar, the fire control unit can be put on in bearing and in range by an early-warning radar station.

Visual Target Acquisition, independent picking up of a target by the optical layer who controls the tracker by means of a joy-stick and a telescope. The range is adjusted by the radar operator.

Target acquisition with external optical putter-on (see above).

As soon as the target has been detected the radar beam is locked on to the target and the unit can be switched over to automatic tracking. There are four methods of tracking a target by this system:

1 Automatically by radar.
2 Optically by visual layer.

Super-Fledermaus deployed for action

Super-Fledermaus in the transport position

3 Automatically without radar, with regenerative control.
4 Automatically without radar but with memory.

The radar operates as a micro-wave pulse radar and was designed by Albiswerk, Zurich AG. The search indicator can be switched over automatically or manually to Plan Position Indicator or Range Height Indicator as desired. The range indicator which shows on two simultaneously visible traces both the entire range from 0–40km (A-display) and a greatly expanded zone of ±1km around the target (R-display).

The Super-Fledermaus analogue computer determines the firing data for the guns. From the moment a target is tracked by the fire control radar or the visual tracker, the present position data are continuously fed into the computer, this determines the data and transmits the data to the guns through cables. The computer also allows for the following: individual horizontal and vertical parallaxes for three different gun emplacements, wind speed and wind direction, air density correction and individual muzzle velocities for each gun. Test and calibration instruments are built into the system, this allows a thorough control of the important electrical functions at any time.

Brief Data: The trailer weighs approx. 5000kg and is 4600mm in length, 2300 wide and 2750mm high. Tracker—tracking speeds in bearing are a max of 2000 mils/s and 1000 mils/s in elevation. Radar—works on X-band, minimum measuring range 200m, maximum scanning range 50km, maximum tracking range 40km.

Employment

Used by many countries including India, Italy, Japan, Switzerland and West Germany. It is built under licence in India, Italy and Japan. It can be used in conjunction with 35mm or 40mm A/A guns.

TACFIRE

Artillery Fire Direction System

TACFIRE is a total command and control system for the field artillery and not only encompasses the battalion FDC and its sub-elements, but adds a division FDC and a Fire Support Co-ordination Element as required for larger Army operations. The requirement for TACFIRE was written in 1966 and in 1967 the contract was awarded to Litton Industries (Data Systems Division).

All the equipment is housed in man-transportable transit cases and can be mounted in S-280 shelters for deployment. TACFIRE improves the accuracy, capability and time responses of the field artillery.

The forward observer (FO) enters all the relevant target characteristics on the Fixed Format Message Entry Device (FFMED), the co-ordinates and target size etc. are entered on the thirty-16 character switches, this data is then transmitted on a 1·3 second digital burst direct to the computer at the FDC. At the FDC the Data Terminal automatically checks the message for errors and corrects any errors made, an acknowledgement signal is then sent to the forward observer. The message is then stored in the AN/GYK-12 (L-3050) computer, the RAM (Random Access Memory) stores 196,608 32-bit words of memory and the MLU (Memory Loading Unit) stores an additional two million eight-bit characters. The computer also stores Met. reports, tactical data, front line positions, ammunition and fire unit status, weapon availability, ranges, rates of fire, muzzle velocities, in fact all the data required.

After the message is received the computer Tactical Fire Control Programme processes the target information and using the weapon data processed by the Ammunition and Fire Unit Status programme selects the weapon, shell, number of rounds required to defeat the target etc. The normal adjusting rounds are simulated on the computer. If the target is out of range or too large for the guns available the computer will generate a request for additional fire message. Or if the target is in a safety area, a warning message is given.

The TACFIRE computer also calculates

The Fixed Format Message Entry Device

all the gun data within some 7seconds from the forward observers first message. The FDO then has all the data he requires to decide how to counter the enemy threat.

The Data Display Group provides the information to the FDO. The Digital Plotter Map automatically plots the latest information on a geographical display using the Electronic Tactical Display. The Electronic Line Printer prints at the rate of 500wpm details of the fire mission request and the fire mission processing results, so that the FDO can review the computers recommended solution.

If the FDO wants to modify the solution he instructs the computer via the Artillery Control Consul (ACC). The ACC operator presses the transmit button and the selected fire command is then sent to the batteries using the Data Terminal.

The message is printed out on the Battery Display Unit (BDU). The battery officer orders the commencement of fire and reports this to the FDO, he in turn informs the FO. The FO observes the results and if any adjustment is required transmits his requirement via the FFMED. After the target is defeated the FO informs the computer via the FFMED and the computer updates its ammunition and target information etc.

The reliability of TACFIRE elements ranges from 1000–32,000 hours MTBF. If a failure occurs it can be readily detected and isolated to a few cards through the use of resident computer programmes. Operating personnel can locate a faulty card, replace it and restore the system to normal operation in less than 10 minutes.

There is also the Electronic Tactical Display (ETD) at the division FDC to analyse intelligence data. Another feature of TACFIRE is that the survey and Met. information can also be fed into the system. TACFIRE can also perform the job of Preliminary Target Analysis, i.e. the capability of artillery to defeat the target, or should aircraft or armour be used.

Employment

Trials with the United States Army.

ABOVE
Electronic Tactical Display

TOP RIGHT
Variable Format Message Entry Device

RIGHT
Artillery Control Console

A Note on Soviet Artillery Fire Control Equipment

The Soviets use a wide range of radar equipment both for the location and control of anti-aircraft fire and for use against ground targets. Information on these is conflicting, some reports give the NATO adopted codename, whilst others give the Russian designation, some even have a 'mix' of the names. The list below should therefore be taken as provisional.

1 *Anti-Aircraft Fire Control Equipment*

SON-4 (NATO designation Fire-Can)

This replaced *Whiff* (NATO codename). SON-4 is an S-band Fire Control System for medium anti-aircraft guns, i.e. 57mm and 85mm. It is believed to have been developed from World War II American Radar SCR-584. The *Fire-Wheel* (NATO codename) Fire Control Radar was developed from the *Fire-Can*.

SON-9 and SON-9a Fire Control Radar

Used to control the 57mm S-60 anti-aircraft gun, used in conjunction with the PUAZO-6/60 director.

SON-30 Fire Control Radar

This is used to control the 130mm KS-30 heavy anti-aircraft gun. Another source gives the SON-30 the name *Flap-Wheel* (NATO).

Directors

The PUAZO-6 replaced the PUAZO-3 and the following types are used:
PUAZO-6/60 for the 57mm Anti-Aircraft Gun S-60
PUAZO-6/12 for the 85mm Anti-Aircraft Gun KS-12
PUAZO-6/18 for the 85mm Anti-Aircraft Gun KS-18
PUAZO-6/19 for the 100mm Anti-Aircraft Gun KS-19
PUAZO-6/30 for the 130mm Anti-Aircraft Gun KS-30

2 *Battlefield Radars*

The Soviet 122mm M-1963 (D-30), 130mm M-1954, 152mm M-1955 (D-20) and 203mm M-1955 weapons are controlled either by:

Small Yawn (NATO codename)

Mobile BSR on a tracked vehicle can be used for mortar location or counter battery roles.

Porktrough (NATO codename)

Mobile BSR mounted on a tracked vehicle, for example the AT-L, used for counter battery roles.

SNAR-2

This is the oldest and is mounted on an AT-L tracked vehicle. Has X-band and orange-peel antenna. Can direct or detect artillery at ranges of 9000m. Settings for surveillance are 1, 5, 10 or 40km.

M-18 FADAC — United States

This is a gun direction computer and is used to compute firing data for artillery weapons from data inputs defining target location, weapon location and prevailing conditions of equipment, material and weather. It is no longer in production.

AN/GVS-3 — United States

This is a laser range finder and weighs approx 14kg. It has an accuracy of ±5m in range, ±2m in azimuth and ±2m in elevation. It has an operating range of 250 to 10,000m.

GR-8 — United States

This is a Sound Ranging Set and weighs approx 46kg and has an accuracy of 0–150m. It locates mortars and artillery, and is a passive device and does not require line of sight.

SIAG — United States

The Surveying Instrument, Azimuth Gyroscopic will replace the current ABLE orientor used by the United States Army. SIAG weighs 18kg and is accurate to 0·3mils at mid latitudes.

AN/UMQ-7 — United States

This is an Artillery Meteorological Sounding System and provides upper atmospheric data for the artillery. The components of the system include:

- 45kw Diesel generator
- Rocket launcher
- Hydrogen generator AN/ML-536
- Inflation and launching device AN/ML-594
- Atmospheric meteorological probe AN/AMQ-22 and AN/AMQ-23
- Balloon AN/ML-566
- Ancillary shelter AN/TMQ-26
- Communications shelter
- Backup meteorological observation set
- Automatic atmospheric sounding set AN/TMQ-19

The data is acquired, processed and computed automatically, and the various meteorological messages are made available to field commanders within minutes after completion of a sounding. The system is mobile (i.e. 4×4 trucks) and it has a range of 160km with tracking accuracies of ±16m in range and ±0·05m in elevation and bearing.

AN/USQ-56 — United States

This is a long range position determining system that is currently under development (LRPDS).

Location Radars — United States

AN/MPQ-4
AN/MPQ-10A
AN/MPQ-501
AN/TTQ-28
KPQ/1

LPD-20 — Italy

The LPD-20 Search Radar has been developed by Contraves Italiana of Rome. It is mounted on a two-wheeled trailer although a version has also been developed to fit on the rear of a truck. It is reported to have a detection range of some 20km.

DOMTI — Sweden

The DOMTI Search Radar is built by Telefonaktiebolaget L. M. Ericsson of Sweden. It has a range of approx 20km.

Bibliography

Listed below are some of the publications that will enable the reader to develop his knowledge of artillery. Some of these have only sections on artillery. Without doubt the best two books are *Barrage—The Guns In Action* and *The Guns 1939–1945*, both by Ian V. Hogg. These give the reader full descriptions of the differences between a gun and howitzer, types of shell, how a gun works, development of all types of artillery and recoilless weapons up to the end of 1945, and the employment of artillery by all the powers during World War II.

Armoured Fighting Vehicles of the World, Christopher F. Foss, Ian Allan, Shepperton, England, 1971.
The Arms Trade With the Third World, SIPRI, Stockholm, Sweden, 1971.
Barrage—The Guns In Action, Ian V. Hogg, Macdonald, London, England, 1971.
Design and Development of Weapons, History of the Second World War, HMSO/Longman, London, 1964.
Dictionary of Modern War, Edward Luttwak, Allen Lane, The Penguin Press, London, 1971.
Die Armeen Der NATO, Fremde Heere, Friedrich Wiener, J. F. Lehmanns, München.
Die Armeen Der Neutralen Und Blockfreien Staaten Europas, Friedrich Wiener, J. F. Lehmanns, München, 1969.
Die Armeen Der Ostblock—Staaten, Friedrich Wiener, J. F. Lehmanns, München.
Die Deutschen Geschütze 1939–1945, F. M. von Senger und Etterlin, J. F. Lehmanns, München, 1960.
Die Streitkräfte Der Siebziger Jahre, Friedrich Wiener, J. F. Lehmanns, München, 1971.
Footslogger's Cannon, Colonel J. Crossman, Ordnance July/August, 1967.
The Guns 1939–1945, Ian V. Hogg, Macdonald, London, 1969.
In The Soviet Arsenal, Leo Heiman, Ordnance January/February, 1968.
Jane's Weapons Systems, published yearly by Sampson and Low, London.
The Long Tom Story, Konrad F. Schreier Jnr, Ordnance November/December, 1967.
New Trends In Artillery, Gilbert J. Melow Jnr, Ordnance Magazine.
A Pictorial Arsenal of America's Combat Weapons, Will Eisner, Sterling Publishing Company Inc, New York, 1962.
The United States Army in World War II—The Technical Services:
The Ordnance Department—Procurement and Supply
The Ordnance Department—Planning Munitions For War
Waffen Und Gerät Der Sowjetischen Landstreitkräfte, Richard F. Arndt, Walhalla U. Praetoria, Germany, 1971.
War Material Used by Vietcong in South Vietnam, Vol 2, 1963.
War Material Presumably In Service in North Vietnam, Volume 1, 1963.
Weapons of World War Two, G. M. Barnes, Van Nostrand, New York, 1947.

Photo credits

The photographs used to illustrate this book have been received from many governments, companies and individuals all over the world. The sources, where known, are listed below. Should however, any of these photographs be wrongly credited, the author would be pleased to know.

Bennett, R. M. 130, 131
British Manufacture & Research Company 45, 46
Canadian Armed Forces 137
Central Office of Information 43 (foot), 44
Central Press Photographs Limited 36, 37
Contraves AG, Zurich, Switzerland 179, 180, 181, 182
DAF, Holland 114 (top)
EMI Limited 170
Ferranti Limited (EDED) 171
Finnish Ministry of Defence 17, 19, 150 (centre)
Foss, Christopher F. 34, 41 (top), 49 (top), 74, 112, 173, 174
French Army 20, 22, 125 (foot), 153
General Electric Company (Armament Dept), USA 135
German Army Public Relations 2, 29 (top), 31, 48, 125 (centre), 154
Gesellschaft für ungelenkte Flugkörpersystem MBH 155
N.V. Hollandse Signaalapparaten 175
Indian Government 39
Imperial War Museum 33, 38, 41 (foot), 42, 43 (top), 85
Israel Defence Forces (General Staff) 56, 64 (foot), 76, 162 (foot)
Litton Systems Incorporated (Data Systems Div) 183, 184
Lockheed Aircraft Service Company 122
Marconi Space and Defence Systems Limited 169, 172
Moggridge, C. W. 49 (centre), 102 (top)
Nissan Motor Company Limited, Japan 157 (top)
Norwegian Army 125 (top)
Oerlikon 105, 107, 109, 110
Pegaso, Spain 117 (centre)
Radartronic A/S 167
Republic of Vietnam 147
Rheinmetall GMBH 29 (foot), 32 (foot), 117 (foot)
Rover Company Limited 35
Saab-Scania (Aerospace Division) 178
Simonsen Radio, Norway 177
Steyr 99 (foot), 128
Swedish Army (Army Material Department) 28, 95, 96, 97, 98, 99 (top), 100, 102 (centre and foot), 103, 104
TASS 13, 62 (foot), 64 (top), 66 (top), 68, 150 (top), 161
Thomson-Brandt (Armament Department) 23, 24, 25, 26, 27
Thomson-CSF, Division Radar de Surface 168
United States Army 115, 120, 121, 123, 127, 128, 129, 132, 133, 138, 139, 141, 159
United States Marine Corps 119 (top), 152 (foot)
Westland Helicopters Limited 49 (foot)
J. I. Taibo, Spain 93, 101, 108, 111, 118, 119 (centre), 152 (top), 157 (foot), 158 (top)
Susuma Yamada, Japan 112 (foot), 114 (centre and foot), 117 (top), 119 (foot), 151 (foot)

Index

The Index has been divided into five Sections: Equipment, Guns, Mortars, Multiple Rocket Systems and Vehicles

Equipment

Guns

Mortars

Multiple Rocket Systems

Vehicles (Tractors and SPGS)